Operator Approximant Problems Arising from Quantum Theory

 Birkhäuser

Philip J. Maher

Operator Approximant Problems Arising from Quantum Theory

 Birkhäuser

Philip J. Maher
London, United Kingdom

ISBN 978-3-319-87019-9   ISBN 978-3-319-61170-9   (eBook)
DOI 10.1007/978-3-319-61170-9

Mathematics Subject Classification (2010): 34-XX, 46-XX, 46N50, 41-XX

Printed on acid-free paper

This book is published under the trade name Birkhäuser, www.birkhauser-science.com
The registered company is Springer International Publishing AG
The registered company address is: Gewerbestrasse 11, 6330 Cham, Switzerland

To the memory of my mother, Marjorie Rose (1914–2012); and to the future of Shirley and Anna.

Preface

This book represents an account of some parts of operator theory, developed mainly since the 1980s whose problems have their roots in quantum theory. The research presented is in non-commutative operator approximation theory or, to use Halmos' terminology, in operator approximants. The crucial concept of approximant is explicated in Chap. 1 ("What this book is about") where the range of problems is outlined. The setting is mainly, but not exclusively, the Von Neumann-Schatten classes $\mathcal{C}_p$.

Thus, quantum chemistry—approximating a Hamiltonian and the Lowdin orthogonalization—precipitates Chap. 3 ("Self-adjoint and positive approximants") and Chap. 6 ("Unitary, isometric and partially isometric approximants") respectively. The commutation relation of quantum mechanics precipitates Chap. 4 ("Commutator approximants").

I have tried—by some necessary simplification—to present the quantum theory background as self-contained. The book therefore assumes no scientific knowledge on the part of the reader. In any case, if the reader is interested in the mathematics alone then she or he could skip the quantum theory motivational sections; but that would be to miss some of the interest.

Obviously, this book presents the necessary mathematical machinery to tackle the various approximant problems. Specifically, Chap. 2 states the Aiken, Erdos and Goldstein result in differentiating the map $\mathcal{C}_p \to \mathbb{R}^+$ given by $X \mapsto \|X\|_p^p$, crucial for most of this work; and Chap. 5 develops the material on spectral approximants required in Chap. 6.

Chapter 2 onward come equipped with a set of exercises whose purpose is to extend, in various directions, the material presented in the body of the chapter. I strongly advise the reader to tackle these exercises since they will, as it were, enable the reader to actively participate in the content of this work. Solutions of many of the exercises can be found in the various papers discussed in the "Notes" with which each chapter concludes.

The reader of this book is expected to have a background in Hilbert space operator theory approaching that of Halmos' marvellous "A Hilbert Space Problem

Book" [23] (to which reference is frequently made). For such a reader this book is suitable for study at postgraduate level.

I thank Dr. Rehana Bari for producing the book in immaculate LaTeX.

I thank Dr. Roger Schafir for discussions on quantum theory.

I thank Dr. Thomas Hempfling of Birkhäuser for the forbearance he has shown over the long gestation of the typescript.

I thank Dr. Matthew M. Jones and Dr. Thomas Bending for their characteristically meticulous checking of the final manuscript which saved me from making a lot of foolish errors.

Some of this work originated, long ago, in my Ph.D thesis. I thank Dr. John Erdos for the help he so freely and generously gave me as my Ph.D supervisor.

London, UK   Philip J. Maher
March 2016

Contents

Chapter 1
What This Book Is About: Approximants

The key concept of this book is that of an approximant (the characteristically snappy term is due to Halmos [21]). Let $\mathcal{L}$, say, be a space of mathematical objects (complex numbers or square matrices, say); let $\mathcal{N}$ be a subset of $\mathcal{L}$ each of whose elements have some "nice" property p (of being real or being self-adjoint, say); and let A be some given, not nice element of $\mathcal{L}$; then a p-approximant of A is a nice mathematical object that is nearest, with respect to some norm, to A. In the first example just mentioned, a given complex number z has its real part $\mathcal{R}z(= \frac{z+\bar{z}}{2})$ as its (unique) real approximant. In the second example, a given square matrix A has (by Theorem 3.2.1) its real part $\mathcal{R}A(= \frac{A+A^*}{2})$ as its unique self-adjoint approximant.

An approximant then minimizes the distance between the set $\mathcal{N}$ of nice mathematical objects in $\mathcal{L}$ and the given object A. Thus, with $\|\cdot\|$ the norm on $\mathcal{L}$, an element A_0 in $\mathcal{N}$ is a p-approximant of A if, for all X in $\mathcal{N}$,

$$\|A - A_0\| \leq \|A - X\|.$$

The concept of an approximant is illustrated by the following diagrams :
Figure 1.1 shows A_0 as the unique approximant of A; Fig. 1.2 shows every point on the inner circle as an approximant of A. There are, thus, in each context several problems:

(I) find an approximant of the given object A;
(II) decide if it is unique.

Figures 1.1 and 1.2 suggest that the uniqueness of an approximant depends—in part, at least—on the "shape" of the set $\mathcal{N}$ of nice objects, as will be confirmed in the context of the Von Neumann-Schatten classes $\mathcal{C}_p$ and norms $\|\cdot\|_p$, for $1 < p < \infty$ (See Theorem 2.4.1). Notice that in the two examples mentioned earlier (of real approximation and of self-adjoint approximation), where there is a unique approximant the set $\mathcal{N}$ (of real numbers and of self-adjoint matrices) is convex in each case.

© Springer International Publishing AG 2017
P.J. Maher, *Operator Approximant Problems Arising from Quantum Theory*,
DOI 10.1007/978-3-319-61170-9_1

Fig. 1.1 This shows A_0 as
the unique approximant of A

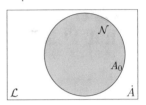

Fig. 1.2 This shows every
point on the inner circle as an
approximant of A

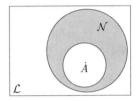

In this work, the space $\mathcal{L} = \mathcal{L}(H)$, the space of all bounded, linear operators on a separable Hilbert Space H. The set $\mathcal{N}$ consists, variously, of:

(a) self-adjoint and positive operators (Chap. 3);
(b) commutators $AX - XA$, generalized commutators $AX - XB$ and self-commutators $X^*X - XX^*$, for varying X (Chap. 4);
(c) normal operators each of whose spectrum is contained in some prescribed closed set in the complex plane (Chap. 5);
(d) and unitary, isometric and partially isometric operators (Chap. 6).

A substantial part of the work here deals with approximant problems with respect to the Von Neumann-Schatten norms $\| \cdot \|_p$, where $1 \le p < \infty$, defined on the Von Neumann-Schatten classes $\mathcal{C}_p$ of operators. With A fixed in $\mathcal{L}(H)$, it is necessary that in minimizing the quantity

$$\|A - X\|_p, \ 1 \le p < \infty,$$

the varying operator X in $\mathcal{N}$ be such that, additionally, $A - X \in \mathcal{C}_p$.

The Von Neumann-Schatten norms play a significant role because of their local properties. Convexity considerations can be used to show that the map

$$X \mapsto \|X\|_p^p, \ 1 < p < \infty,$$

(from $\mathcal{C}_p$ to $\mathbb{R}^+$) is differentiable; and a powerful result of Aiken, Erdos and Goldstein, Theorem 2.4.4 [1, Theorem 2.1] gives an explicit formula for the derivative of this map. This suggests a third general problem be added to the two posited earlier, viz.

Fig. 1.3 Here, the set of
global minimizers is a subset
of the set of critical points

(III) classify the critical points of the map

$$F_p : X \mapsto \|A - X\|_p^p, \ 1 < p < \infty,$$

where $A - X \in \mathcal{C}_p$, $X \in \mathcal{C}_p$ and $X \in \mathcal{N}$.

Although (the global) problems (I) and (II) and (the local) problem (III) can, in some cases, be solved independently of each other (as in the case of self-adjoint approximation, for instance), there is a connection between global and the local considerations. For if V is a global minimizer of the map

$$F_p : X \mapsto \|A - X\|_p^p, \ 1 \leq p < \infty$$

then, for $1 < p < \infty$, it is a critical point of it (Fig. 1.3). Thus, for $1 < p < \infty$, the set of global minimizers of F_p is a subset of the set of critical points of it:

(The inclusions may not always be strict: in some cases every critical point may be a global minimizer.) Thus, in searching for a global minimizer of F_p we can restrict attention to its critical points. Therefore, solving problem (III) can be a step towards solving problems (I) and (II). This approach is adopted in tackling the problem of finding partially isometric approximants (Chap. 6).

Chapter 2
Preliminaries

This chapter presents, and highlights, material (much of which will be familiar to the reader) required for the rest of this work.

2.1 Operators (in General)

Throughout, H will denote a complex, separable Hilbert space with inner product $\langle,\rangle$ and norm $\|\cdot\|$. $\mathcal{L}(H)$ will denote the space of bounded, linear operators mapping H onto H (here, bounded means that the supremum norm $\|\cdot\|$, defined by $\|A\| = \sup_{\|f\|\le 1}\|Af\|$, is finite; occasionally, we shall write $\|\cdot\|_\infty$ for the supremum norm $\|\cdot\|$).

With each A in $\mathcal{L}(H)$ we associate two subspaces of H denoted $\mathrm{Ker}A$ and $\mathrm{Ran}A$ and defined by $\mathrm{Ker}A = \{f: Af = 0 \text{ where } f \in H\}$ and $\mathrm{Ran}A = \{g: g = Af \text{ where } f \in H\}$. Observe that $\mathrm{Ker}A$ is closed but $\mathrm{Ran}A$ is not necessarily closed.

If $A \in \mathcal{L}(H)$ then A^* denotes the adjoint of A. The operator A is **normal** if $AA^* = A^*A$ and **self-adjoint** if $A = A^*$. Note that A is self-adjoint if and only if $\langle Af, f\rangle$ is real for all f in H. If $\langle Af, f\rangle \ge 0$ for all f in H, then A is **positive**; this is denoted by $A \ge 0$. If $A \ge 0$ and $\mathrm{Ker}A = \{0\}$ then A is **strictly positive**; this is denoted by $A > 0$. Observe that for every A in $\mathcal{L}(H)$ we automatically have $A^*A \ge 0$.

The projections form a subset of the positive operators. An operator P is called a **projection** (or orthogonal projection) if $P = P^* = P^2$; equivalently, P is a projection if $Pf = f$ for all f in $(\mathrm{Ker}P)^\perp$.

An **isometry** U is an operator such that $\|Uf\| = \|f\|$ for all f in H; equivalently, U is an isometry if and only if $U^*U = I$. If U^* is an isometry (so that $UU^* = I$) then U is called a **co-isometry**. An invertible isometry is called a **unitary** operator; equivalently, U is unitary if and only if $U^*U = UU^* = I$.

A **partial isometry** U is an operator that is isometric on the orthogonal complement of its kernel: thus U is partial isometry if $\|Uf\| = \|f\|$ for all f in

© Springer International Publishing AG 2017
P.J. Maher, *Operator Approximant Problems Arising from Quantum Theory*,
DOI 10.1007/978-3-319-61170-9_2

$(\mathrm{Ker}U)^{\perp}$. For a partial isometry U the operators U^*U and UU^* are, respectively, the projections onto $(\mathrm{Ker}U)^{\perp}$ and $\mathrm{Ran}U$ (called, respectively, the **initial space** of U and the **final space** of U). For a partial isometry U we shall write $E_U = U^*U$ and $F_U = UU^*$ (so that $E_{U^*} = F_U$). Thus, a partial isometry U is normal if and only if $E_U = F_U$, that is, if and only if its initial and final spaces coincide. Note also that an operator U is a partial isometry if and only if $U = UU^*U$.

Recall that the **spectrum**, $\sigma(A)$, of A is defined as follows: $\sigma(A) = \{z\colon (A - zI)$ is non-invertible$\}$. The **spectral radius**, $r(A)$, of A is defined as $r(A) = \sup\{|z| : z \in \sigma(A)\}$. It can be shown that $r(A) = \lim_{n\to\infty} \|A^n\|^{\frac{1}{n}}$. For a normal operator A it follows easily that $r(A) = \|A\|$. An operator A is **quasinilpotent** if $\lim_{n\to\infty} \|A^n\|^{\frac{1}{n}} = 0$, that is, if $\sigma(A) = \{0\}$. Thus, a normal quasinilpotent operator is the zero operator.

The **numerical range**, $W(A)$, of an operator A is defined by $W(A) = \{\langle Af, f \rangle : \|f\| = 1\}$ and its **numerical radius**, $w(A)$, is defined by $w(A) = \sup\{|z| : z \in W(A)\}$. Recall that a set S is **convex** if whenever $f \in S$ and $g \in S$ then $\alpha f + (1-\alpha)g \in S$ for all $0 < \alpha < 1$. It can be shown that the numerical range of an operator is convex [23, Problem 210]. Further, the closure of the numerical range of a normal operator is the **convex hull** of its spectrum [23, Problem 216]. (Recall that the convex hull of a set S, in symbols $\mathrm{conv}S$, is the smallest convex set that includes S). Thus if A is normal

$$\overline{W(A)} = \mathrm{conv}\,\sigma(A). \tag{2.1.1}$$

Observe that $w(A) \leq \|A\|$ for all A, equality holding if A is normal.

2.2 The Spectral Theorem and the Polar Decomposition

The spectral theorem is the infinite-dimensional generalization of the result from linear algebra that a finite, square normal matrix may be diagonalized. The **spectral theorem** for normal operators says that a normal operator is unitarily equivalent to a multiplication. Explicitly, if A is a normal operator in $\mathcal{L}(H)$ then there exists a bounded, measurable function ϕ on some measure space X with measure μ and an isometry U on $L_2(\mu)$ onto H such that

$$(U^{-1}AUf)(x) = \phi(x)f(x)$$

for each f in $L_2(\mu)$ [20, 39, Theorem 1.6]. The spectral theorem can be expressed in terms of spectral measures as follows: if A is a normal operator then there exists a unique spectral measure $P(\cdot)$ such that

$$\langle Au, v \rangle = \int_{\sigma(A)} z d \langle P(z)u, v \rangle$$

[20, 39, Theorem 1.12] (to avoid confusion we are writing here u, v for vectors of H rather than the usual f, g).

The spectral theorem has many consequences. One is that every normal operator is the norm limit of diagonal ones. Another is that for a positive operator Z

$$\operatorname{Ker} Z^r = \operatorname{Ker} Z, \quad \text{if} \quad 0 < r < \infty. \tag{2.2.1}$$

A third consequence is **Fuglede's Theorem:** this says that if A is normal and $AT = TA$, for some T, then $A^*T = TA^*$ [23, Problem 192].

From the spectral theorem one may deduce the **functional calculus**. This gives a meaning to $f(A)$ when f is not a polynomial. The functional calculus says that if A is a normal operator and f is a bounded, Borel measurable function on $\sigma(A)$ then the map $f \rightarrow f(A)$, from the algebra of bounded, Borel measurable functions on $\sigma(A)$ to $\mathcal{L}(H)$, defined by

$$\langle f(A)u, v \rangle = \int_{\sigma(A)} f(z) d \langle P(z)u, v \rangle$$

is an algebra homomorphism; so that, e.g.,

$$\| f(A)u \|^2 = \int_{\sigma(A)} |f(z)|^2 d \langle P(z)u, u \rangle \tag{2.2.2}$$

[20, 39, Theorem 1.22], cf. [40, Theorem 1.7.7]. The functional calculus embodies the result that, for normal A, if the sequence $\{p_i\}$ converges uniformly to a bounded, Borel measurable function f on $\sigma(A)$ then $p_i(A)$ converges strongly to $f(A)$:

$$p_i \rightarrow f \text{ uniformly on } \sigma(A) \implies \| p_i(A) - f(A) \| \rightarrow 0;$$

whence it follows:

$$p_i(A)B = Cp_i(A) \text{ and } p_i \rightarrow f \text{ uniformly on } \sigma(A) \implies f(A)B = Cf(A). \tag{2.2.3}$$

It follows from the functional calculus that every positive operator T, say, has a unique positive square root, denoted $T^{1/2}$. For arbitrary A in $\mathcal{L}(H)$ we denote the unique, positive square root of (the positive operator) A^*A by $|A|$; thus, $|A| = (A^*A)^{1/2}$. The operator $|A|$ is called the **modulus** of A.

The analogy of the polar decomposition of a complex number ($z = e^{i\theta} |z|$ for z in $\mathbb{C}$) is the polar decomposition of an operator. We state this result, and collect some of its variants, below.

Theorem 2.2.1 (Polar Decomposition [23, Chap. 16])

(a) Each A in $\mathcal{L}(H)$ may be expressed uniquely as

$$A = U|A|$$

where U is the partial isometry such that $\text{Ker} U = \text{Ker} |A|$ *(and where* $\text{Ran} U = \text{Ran} A$ *);*

(b) *$U^*A = |A|$;*

(c) *further, there exists V in $\mathcal{L}(H)$ such that either V or V^* is isometric and such that V coincides with U on* $(\text{Ker} |A|)^{\perp}$ [23, Solution to Problem 135];

(d) *if, further* $\dim\text{Ker} A = \dim\text{Ker} A^*$ *(equivalently,* $\dim\text{Ker} U = \dim(\text{Ran} U)^{\perp}$*) then the partial isometry U can be extended to a unitary $\hat{U}$ which agrees with U on* $(\text{Ker} |A|)^{\perp}$ *and which can be any isometry mapping $\text{Ker} |A|$ onto* $(\text{Ran} A)^{\perp}$ *cf.* [24, p. 586];

(e) *if A is self-adjoint then the partial isometry U is self-adjoint and commutes with* $|A|$.

(Observe that the condition of (d), $\dim\text{Ker} A = \dim\text{Ker} A^*$ is automatically met in finite dimensions.)

2.3 Compact Operators

There are various, equivalent, definitions of compactness [23, Problem 170]. Here is one. An operator A is called **compact** (or completely continuous) if every sequence $\{f_n\}$ of vectors such that $\|f_n\|$ is bounded (that is, such that $\sup\{\|f_n\|: n \in \mathbb{N}\} < \infty$) has a subsequence $\{f_{n_q}\}$ such that $\{Af_{n_q}\}$ converges (in $\|\cdot\|$). The set $\mathcal{C}_\infty$ of all compact operators in $\mathcal{L}(H)$ is a closed, two-sided ideal that is therefore self-adjoint [23, Problem 170]. (Here, "closed" refers to the sup norm topology; "two-sided ideal" means that linear combinations of operators in $\mathcal{C}_\infty$ are in $\mathcal{C}_\infty$ and that products with at least one factor in $\mathcal{C}_\infty$ are in $\mathcal{C}_\infty$; and "self-adjoint" means that if $A \in \mathcal{C}_\infty$ then $A^* \in \mathcal{C}_\infty$.) That every two-sided ideal of $\mathcal{L}(H)$ is self-adjoint follows using the polar decomposition Theorem 2.2.1(b). Thus, $A \in \mathcal{C}_\infty$ if and only if $|A| \in \mathcal{C}_\infty$. The **Calkin algebra** is the algebra $\mathcal{L}(H)/\mathcal{C}_\infty$.

An operator A is of **finite rank** if $\dim\text{Ran} A < \infty$; in this case $\dim\text{Ran} A$ is called the rank of A. The finite rank operators are compact and form the simplest class of compact operators. The rank 1 operator $x \to \langle x, e\rangle f$, for fixed vectors e and f, is denoted $e \otimes f$ and satisfies

$$(e \otimes f)^* = f \otimes e \qquad \text{and} \qquad A(e \otimes f)B = (B^*e) \otimes (Af) \tag{2.3.1}$$

for A and B in $\mathcal{L}(H)$.

The so-called "**Fredholm alternative**" says that if A is compact and λ is a non-zero complex number then either λ is an eigenvalue of A or $\lambda \notin \sigma(A)$ [23, Problem 179]. Further, if A is compact then $\sigma(A)$ is either finite or countably infinite and in the latter case $\sigma(A)$ consists of a complex sequence converging to 0 [40, Theorem 1.8.2].

Compact operators can be expressed as sums (finite or countably infinite) of rank 1 operators. If $\{\mu_i\}$ is a sequence of positive reals, either finite or infinite and converging to 0, and if $\{\phi_i\}$ and $\{\psi_i\}$ are orthonormal sequences then A is compact if and only if

$$A = \sum_{i=1}^{m} \mu_i(\phi_i \otimes \psi_i) \tag{2.3.2}$$

where $m \leq \infty$; and if A is compact the sequence $\{\mu_i\}$ is uniquely determined and consists of eigenvalues of $|A|$, henceforth denoted $s_i(A)$, arranged in decreasing order and counted accordingly to their multiplicities and where $\{\phi_i\}$ is the corresponding sequence of eigenvectors of $|A|$ and $\mathcal{S}\{\psi_i\} = \mathrm{Ran}A$ [40, Theorem 1.9.3]. The eigenvalues $s_i(A)$, of $|A|$, are called the **singular values** of A.

The spectral theorem for normal, compact operators says that each such operator A can be expressed in the form (2.3.2) with $\{\mu_i\}$ the sequence of non-zero eigenvalues of A, arranged in decreasing order of magnitude and counted according to their multiplicities, and with $\psi_i = \phi_i$, the corresponding sequence of eigenvectors of A [40, Theorem 1.9.2].

Thus, if A is a normal, compact operator then there exists a complete, orthonormal basis of H consisting of eigenvectors of A. Further, it can be shown that if A and B are two commuting, normal, compact operators then there exists a complete orthonormal basis of H consisting of eigenvectors of A and B.

2.4   The Von Neumann-Schatten Classes $\mathcal{C}_p$

For a compact operator A let, as usual, $s_i(A)$ be the (positive) eigenvalues of the compact operator $|A|$, arranged in decreasing order and counted according to multiplicity. If, for some p, such that $0 < p < \infty$,

$$\sum_{i=1}^{\infty} s_i(A)^p < \infty$$

we say that A is in the **Von Neumann-Schatten p class** $\mathcal{C}_p$ and write

$$\|A\|_p = \left[\sum_{i=1}^{\infty} s_i(A)^p \right]^{\frac{1}{p}} \tag{2.4.1}$$

(as before, $\mathcal{C}_\infty$ is identified with the ideal of compact operators). If $1 \leq p < \infty$ then it can be shown that $\| \cdot \|_p$ is a norm and that under this norm $\mathcal{C}_p$ is a Banach space [40, Theorem 2.3.8]; if $0 < p < 1$ then $\mathcal{C}_p$ is a metric space with metric d given by $d(A, B) = \sum s_i(A - B)^p$ (thus, if $0 < p < 1$ then $\| \cdot \|_p$ fails to be homogeneous). For all p, such that $0 < p < \infty$, $\mathcal{C}_p$ is a two-sided ideal of $\mathcal{L}(H)$ and $\|A\|_p = \|A^*\|_p$. Thus, if $A \in \mathcal{L}(H)$, $B \in \mathcal{L}(H)$ and $S \in \mathcal{C}_p$ then $ASB \in \mathcal{C}_p$; and it can be proved that, for $1 \leq p < \infty$,

$$\|ASB\|_p \leq \|A\| \|S\|_p \|B\|. \tag{2.4.2}$$

For $0 < p < \infty$, it can be shown that C_p is an increasing function of p whilst $\|\cdot\|_p$ is a decreasing function of p; that is, if $0 < p_1 \leq p_2 < \infty$ then $C_{p_1} \subseteq C_{p_2} \subset C_\infty$ and if $A \in C_{p_1}$ then $\|A\|_{p_1} \geq \|A\|_{p_2} \geq \|A\|$. Further, it can be shown that, for finite rank X, the function $p \mapsto \|X\|_p$ is a continuous function of p, where $1 \leq p < \infty$.

The class C_1 is called the **trace class**. If $A \in C_1$ and if $\{\phi_i\}$ is any complete orthonormal basis of H, then the quantity $\tau(A)$, called the **trace** of A and defined by

$$\tau(A) = \sum_{i=1}^{\infty} \langle A\phi_i, \phi_i \rangle,$$

can be shown to be finite and independent of the particular basis chosen [40, Lemma 2.2.1]. If $A \in C_1$, then $\|A\|_1 \geq |\tau(A)|$ [40, Lemma 2.3.3]. It can be shown that $A \in C_p$ if and only if $|A|^p \in C_1$ where $1 \leq p < \infty$ [40, Lemma 2.3.1]. Thus by (2.4.1), with $1 \leq p < \infty$:

$$\|A\|_p^p = \tau(|A|^p) = \sum_{i=1}^{\infty} \langle |A|^p \phi_i, \phi_i \rangle, \tag{2.4.3}$$

so that if $p = 2$

$$\|A\|_2^2 = \sum_{i=1}^{\infty} \|A\phi_i\|^2 \tag{2.4.4}$$

for every compact orthonormal basis $\{\phi_i\}$ of H.

Algebraically, τ has the following properties: if $A \in C_1$, $B \in C_1$ and $S \in \mathcal{L}(H)$ then $\tau(\alpha A + \beta B) = \alpha \tau(A) + \beta \tau(B)$ (where α and β are scalars); $\tau(A^*) = \overline{\tau(A)}$; $\tau(A) \geq 0$ if $A \geq 0$; and $\tau(AS) = \tau(SA)$ (This last property is sometimes referred to as the **invariance of trace**) [40, Theorem 2.2.4].

Every finite rank operator is in C_p, where $0 < p < \infty$, and, for $1 \leq p < \infty$, the set of all finite rank operators is dense in C_p [40, Theorem 2.3.8]. It is easy to check that $\tau(e \otimes f) = \langle f, e \rangle$ cf. [40, p. 90] and hence, by (2.3.1), for A and B in $\mathcal{L}(H)$

$$\tau(A(e \otimes f)B) = \langle BAf, e \rangle. \tag{2.4.5}$$

If A is a normal, compact operator and $\{\alpha_i\}$ is the sequence of non-zero eigenvalues of A (arranged in order of decreasing magnitude and counted according to multiplicity) then $\{|\alpha_i|\}$ is the sequence of non-zero eigenvalues of $|A|$; thus, the normal, compact operator A is in C_p, where $1 \leq p < \infty$, if and only if $\sum_i |\alpha_i|^p < \infty$, and when the normal $A \in C_p$ then [40, pp. 86–87]

$$\|A\|_p^p = \sum_{i=1}^{\infty} |\alpha_i|^p. \tag{2.4.6}$$

An inequality we need for minimization problems goes as follows. If $A \in C_p$, where $1 \le p < \infty$, then

$$\|A\|_p^p \ge \sum_{i=1}^{\infty} |\langle A\phi_i, \phi_i \rangle|^p \tag{2.4.7}$$

for every complete orthonormal basis $\{\phi_i\}$ of H cf. [40, Lemma 2.3.4, Corollary 2.3.6].

We shall appeal to the following equality about operators with orthogonal ranges/co-ranges. It says that if $A + B \in C_p$ where $0 < p < \infty$, if Ran$A \perp$ Ran$B$ and Ran$A^* \perp$ RanB^* then $A \in C_p$, $B \in C_p$ and

$$\|A + B\|_p^p = \|A\|_p^p + \|B\|_p^p \tag{2.4.8}$$

[29, Theorem 1.7(d)]: see Exercise 3.

The class C_2, called the **Hilbert–Schmidt** class, is significant amongst the C_p classes in that it, alone, is a Hilbert space under the inner product $\langle, \rangle$ defined, for each A and B in C_2, by $\langle A, B \rangle = \tau(B^*A)$, the norm derived from this inner product being the $\| \cdot \|_2$ norm; further C_2 is separable since if $\{\phi_i : i \in I\}$ and $\{\psi_j : j \in J\}$ are complete orthonormal bases of H, then $\{\phi_i \otimes \psi_j : i, j \in I\}$ is a complete orthonormal basis of C_2 [40, Theorem 2.4.2]. It follows that each Banach space C_p for $1 \le p < 2$ is separable.

Let p and q be such that $1 \le p, q < \infty$ and $\dfrac{1}{p} + \dfrac{1}{q} = 1$ (when p and q are called **conjugate**); then the dual of C_p, denoted by $(C_p)^*$, is given by $(C_p)^* = C_q$ with $(C_1)^* = \mathcal{L}(H)$ [40, Theorem 2.3.12]. Of course, $(C_2)^* = C_2$.

The convexity of sets of operators is crucial in minimization problems. A set S of operators is **convex** if whenever $X \in S$ and $Y \in S$ then $\alpha X + (1 - \alpha) Y \in S$ where $0 < \alpha < 1$. A norm $\|\| \cdot \|\|$, say, is **strictly convex** if whenever $\|\|X + Y\|\| = \|\|X\|\| + \|\|Y\|\|$ then $aX = bY$ for non-negative reals a and b such that $a + b > 0$. McCarthy [37, Theorem 2.4] proved that the $\| \cdot \|_p$ norm is strictly convex for $1 < p < \infty$. This yields the following result.

Theorem 2.4.1 ([31, Lemma 2.5]) *Let S be a convex set of operators and let X vary in S and be such that, for A fixed, $A - X \in C_p$ where $1 < p < \infty$. Then there is at most one minimizer of $\|A - X\|_p$.*

A normed linear space $\mathcal{C}$, with norm $\|\|\cdot\|\|$, is **uniformly convex** if to each $\epsilon > 0$ corresponds $\delta(\epsilon) > 0$ such that, for X and Y in $\mathcal{C}$, $\|\|\frac{X+Y}{2}\|\| \leq 1 - \delta(\epsilon)$ where $\|\|X\|\| \leq 1$, $\|\|Y\|\| \leq 1$ and $\|\|X - Y\|\| \geq \epsilon$. McCarthy [37, pp. 258–262] proved that $\mathcal{C}_p$ where $1 < p < \infty$, is uniformly convex.

Definition 2.4.2 Let $\mathcal{B}$ be a Banach space. A map $F : \mathcal{B} \to \mathbb{R}$ is said to be (uniformly) **Fréchet differentiable** at X with derivative denoted $D_X F$ if the following limit, given by

$$(D_X F)(T) = \lim_{h \to 0} \frac{F(X + hT) - F(X)}{h},$$

exists uniformly.

A standard result [12, Theorem 1, p. 36] says that if the dual of $\mathcal{C}$, $\mathcal{C}^*$, is uniformly convex then the map $\mathcal{C} \to \mathbb{R}^+$ given by $X \mapsto \|X\|$ is Fréchet differentiable.

Theorem 2.4.3 *The map $\mathcal{C}_p \to \mathbb{R}^+$ given by $X \mapsto \|X\|_p$ for $1 < p < \infty$, is Fréchet differentiable.*

Proof Let $\mathcal{C} = \mathcal{C}_p$ where $1 < p < \infty$. Since $\mathcal{C}^* = (\mathcal{C}_p)^* = \mathcal{C}_q$, where $\dfrac{1}{p} + \dfrac{1}{q} = 1$ so that $1 < q < \infty$, is uniformly convex it follows that $X \mapsto \|X\|_p$ is Fréchet differentiable.

It follows from Theorem 2.4.3 that the map $X \mapsto \|X\|_p^p$ is Fréchet differentiable for $1 < p < \infty$. Aiken, Erdos and Goldstein [1, Theorem 2.1] found an explicit formula for the derivative of this map which we state below (Theorem 2.4.4). This result underpins much of the rest of this work.

Theorem 2.4.4 *Let the map $F_p : \mathcal{C}_p \to \mathbb{R}^+$ be given by $F_p : X \mapsto \|X\|_p^p$. Then:*

(a) for $1 < p < \infty$, the map F_p is Fréchet differentiable with derivative $D_X F_p$ given by

$$(D_X F_p)(T) = p\mathcal{R}\tau[|X|^{p-1} U^* T] \tag{2.4.9}$$

where $X = U|X|$ is the polar decomposition of X;
(b) for $0 < p \leq 1$, provided the underlying space is finite dimensional, the same result holds at every invertible element X.

Finally, we observe that $\|\cdot\|$ and $\|\cdot\|_p$, for $1 \leq p < \infty$, are examples of unitarily invariant norms. A **unitarily invariant norm** $\|\|\cdot\|\|$ is a norm defined on some two-sided ideal $\mathcal{I}$ of $\mathcal{L}(H)$ such that $\|\|UA\|\| = \|\|AV\|\|$ for all unitary operators U and V. It can be shown that $\|\|A^*\|\| = \|\|A\|\|$.

Exercises

1 Prove that the class $\mathcal{C}_2$ is a Hilbert space with respect to the inner product $\langle , \rangle$ defined by

$$\langle A, B \rangle = \tau \left(B^* A \right)$$

for each A and B in $\mathcal{C}_2$.

2 Prove that:

(a) if $|A|^2 \geq |B|^2$ then $\|A\| \geq \|B\|$;
(b) if, further, A is compact then B is compact and $|A| \geq |B|$;
(c) and if, further, $A \in \mathcal{C}_p$, where $1 < p < \infty$, then $B \in \mathcal{C}_p$ and $\|A\|_p \geq \|B\|_p$.

3 Prove that:

(a) if $\operatorname{Ran}A \perp \operatorname{Ran}B$ (**or** if $\operatorname{Ran}A^* \perp \operatorname{Ran}B^*$) then

$$\|A + B\| \geq \max\{\|A\|, \|B\|\}; \tag{1}$$

(b) if $\operatorname{Ran}A \perp \operatorname{Ran}B$ **and** $\operatorname{Ran}A^* \perp \operatorname{Ran}B^*$ then equality holds in (1);
(c) if $A + B \in \mathcal{C}_p$ and if $\operatorname{Ran}A \perp \operatorname{Ran}B$ (or if $\operatorname{Ran}A^* \perp \operatorname{Ran}B^*$) then $A \in \mathcal{C}_p$ and $B \in \mathcal{C}_p$ for $0 < p < \infty$ and

$$\|A + B\|_p^p \geq \|A\|_p^p + \|B\|_p^p \tag{2}$$

for $2 \leq p < \infty$ with equality in the $p = 2$ case;
(d) and if, in (c), $\operatorname{Ran}A \perp \operatorname{Ran}B$ and $\operatorname{Ran}A^* \perp \operatorname{Ran}B^*$ then equality holds in (2) for $0 < p < \infty$.

4 Prove Theorem 2.4.1.

Notes

Halmos' book [23] is a marvelous compendium of results from single operator theory. For a beautiful account of the Von Neumann–Schatten classes see Ringrose [40, Chap. 2]; see also Dunford and Schwartz [13, Chap. XI, 9]. A thorough account of the geometry of Banach space is given by Diestel [12].

Exercises: Exercise 1 is in Ringrose [40, Theorem 2.4.2]; Exercise 2 is in Maher [31, Lemma 3.1], as are Exercise 3 [29, Theorem 1.7] and Exercise 4 [31, Lemma 2.5].

Chapter 3
Self-Adjoint and Positive Approximants

The subject of operator approximation dates back to the 1950s to the seminal work of Fan and Hoffman [16] who studied, in part, self-adjoint approximation. It was not, however, until the 1970s that the subject seems to have taken of, precipitated by the papers of Halmos [21, 22]. In [21] Halmos coined the term "approximant" and focused mainly on positive approximation.

The simpler topic of self-adjoint approximation has links with quantum chemistry which we first outline below.

3.1 Quantum Chemical Background: Approximating a Hamiltonian

The key notion is that of a Hamiltonian: a Hamiltonian governs a quantum chemical system. A Hamiltonian is, technically speaking, a semi-bounded, self-adjoint operator on some infinite-dimensional Hilbert space. The value $\inf\{|z|: z \in \sigma(H)\}$ of the Hamiltonian H is often an eigenvalue (of algebraic multiplicity 1). This eigenvalue and its (essentially unique) eigenvector are called the **ground state energy** E and the **ground state wave function** ψ of the system. Finding—or rather, approximating to—E and ψ is of great importance in chemistry since the ground state of a system is its most stable state and the one most often found in nature.

Given that the Hamiltonian is an operator on an infinite-dimensional Hilbert space and given that only finite data are available, one seeks to replace the original infinite-dimensional Hilbert space by a large finite-dimensional one and thereby find an approximation to E (and perhaps ψ) by some numerical procedure. In effect, one splits up the original system into smaller, better understood subsystems and then combines them back (linearly perhaps) so as to approximate, or mirror, the original system.

© Springer International Publishing AG 2017
P.J. Maher, *Operator Approximant Problems Arising from Quantum Theory*,
DOI 10.1007/978-3-319-61170-9_3

To illustrate this, suppose molecule A_j consists of N_j atoms ($j = 1, 2$) and these two molecules, A_1 and A_2, are combined back to form molecule A_3 which has N_3 atoms. Thus, one wishes to obtain—or approximate to—the ground state energy and wave function E_{A_3} and ψ_{A_3} of the system A_3 from the simpler ground state energies and wave functions E_{A_j} and ψ_{A_j} of the constituent systems A_j ($j = 1, 2$). For example, the formula cited by Goldstein and Levy [17]

$$C_3H_8 = 2C_2H_6 - CH_4$$

(which reads "one propane equals two ethanes minus one methane") suggests the obvious expression for the ground state energy $E_{C_3H_8}$ of propane in terms of the ground state energies $E_{C_2H_6}$ and E_{CH_4} of ethane and methane:

$$E_{C_3H_8} = 2E_{C_2H_6} - E_{CH_4}; \qquad (3.1.1)$$

and this latter formula, whilst not correct, is known to be very accurate—the relative error between the two sides of (3.1.1) is less than 0.01% [4, 27].

Suppose now that we are given a matrix A that is meant as a finite-dimensional approximation of the Hamiltonian of some given system but which, unlike a Hamiltonian, is not self-adjoint (perhaps because of some numerical error). Then the natural thing to do is to replace A by a self-adjoint matrix that is nearest to A: that is, we require to find a self-adjoint approximant of A [18, p. 712].

3.2 Self-Adjoint Approximants

The result that for a complex number z, $\mathcal{R}z$ is the real approximant of z suggests that, for an operator A, $\mathcal{R}A$ is a self-adjoint approximant of A. This is indeed true and applies to all self-adjoint (i.e. unitarily invariant) norms including the supremum norm $\| \cdot \|$ and the Von-Neumann–Schatten norms $\| \cdot \|_p$, where $1 \leq p < \infty$ [14].

Theorem 3.2.1 *Let A be in $\mathcal{L}(H)$ and $\| \cdot \|$ be a unitarily invariant norm defined on a subset $\mathcal{C}$ of $\mathcal{L}(H)$. Let X vary over the self-adjoint operators and be such that $A - X \in \mathcal{C}$. Then $A - \mathcal{R}A \in \mathcal{C}$ and*

$$\|A - \mathcal{R}A\| \leq \|A - X\|. \qquad (3.2.1)$$

In particular, if $\| \cdot \| = \| \cdot \|_p$ and $\mathcal{C} = \mathcal{C}_p$, where $1 \leq p < \infty$, then $A - \mathcal{R}A \in \mathcal{C}_p$ and

$$\|A - \mathcal{R}A\|_p \leq \|A - X\|_p \qquad (3.2.2)$$

with equality occurring in (3.2.2) if, and for $1 < p < \infty$ only if, $X = \mathcal{R}A$.

Proof With $\mathcal{R}A = \dfrac{A + A^*}{2}$ and $X = X^*$ we have (using the homogeneity and self-adjointness of $\||\cdot\||$)

$$\|A - \mathcal{R}A\| = \||\frac{A - A^*}{2}\|| = \frac{1}{2}\||(A - X) - (A - X)^*\||$$

$$\leq \frac{1}{2}[\||A - X\|| + \||(A - X)^*\||] = \frac{1}{2}.2\|A - X\| = \|A - X\|$$

so that $A - \mathcal{R}A \in \mathcal{C}$ and the inequality (3.2.1) follows.

The inequality (3.2.2) is immediate from (3.2.1) (The $p = \infty$ version is the supremum norm inequality $\|A - \mathcal{R}A\| \leq \|A - X\|$ on "$\mathcal{C}$" $= \mathcal{L}(H)$).

The uniqueness assertion, for $1 < p < \infty$, follows from Theorem 2.4.1 since the set of self-adjoint operators is convex: for if X_1 and X_2 are self-adjoint so, too, is $\alpha X_1 + (1 - \alpha)X_2$ where $0 < \alpha < 1$. $\square$

The following example shows, first, when $p = 1$ an operator A does not have $\mathcal{R}A$ as its unique self-adjoint approximant and, second, when $0 < p < 1$, the inequality (3.2.2) of Theorem 3.2.1 breaks down.

Example 3.2.2 Let $H = \mathbb{C}^2$ and let A and I be given by

$$A = \begin{bmatrix} 1 & 1 \\ 0 & 1 \end{bmatrix} \quad \text{and} \quad I = \begin{bmatrix} 1 & 0 \\ 0 & 1 \end{bmatrix}.$$

Then the eigenvalues of $|A - \mathcal{R}A|$ and $|A - I|$ are $\frac{1}{2}, \frac{1}{2}$ and $1, 0$ respectively. So, first $\|A - \mathcal{R}A\|_1 (= 1) = \|A - I\|_1$ yet $I \neq \mathcal{R}A$; and, second, if $p = \frac{1}{2}$ then $\|A - \mathcal{R}A\|_{\frac{1}{2}} = 2 > 1 = \|A - I\|_{\frac{1}{2}}$.

For $p = \infty$ uniqueness also breaks down.

Example 3.2.3 ([14]) Let f_1 and f_2 be two orthonormal vectors, let E_k, where $k = 1, 2$, be the orthogonal projection onto $\mathcal{S}\{f_k\}$ and let $A = E_1 + 2iE_2$ (so that $\mathcal{R}A = E_1$). Then $\|A - \mathcal{R}A\| = \|A\|$.

The local analogue (Theorem 3.2.4 below) of the global result for $\|\cdot\|_p$ (contained in Theorem 3.2.1) shows that, for $1 < p < \infty$, V is a critical point of $X \rightarrow \|A - X\|_p^p$ (where $X = X^*$) if **and only if** it is a global minimizer. The proof is prototypical of some of the more complicated proofs of local results; and may, consequently, be regarded as an introduction to them.

Theorem 3.2.4 *Let A be in $\mathcal{L}(H)$ and $\mathcal{S}$ be defined by*

$$\mathcal{S} = \{X : X \text{ is self–adjoint and } A - X \in \mathcal{C}_p \text{ where } 1 < p < \infty\}.$$

If $S \neq \emptyset$, let $F_p : S \to \mathbb{R}^+$ be defined by

$$F_p : X \mapsto \|A - X\|_p^p.$$

Then V is a critical point of F_p if and only if $V = \mathcal{R}A$.

Proof Let $V = \mathcal{R}A$. Then, by Theorem 3.2.1, V is a global minimizer of F_p and hence, for $1 < p < \infty$, a critical point of F_p.

Conversely, let V be a critical point of $F_p : X \to \|A - X\|_p^p$, where $1 < p < \infty$, and let T be an arbitrary increment of V. Then, by Theorem 2.4.4

$$0 = (D_{A-V}F_p)(-T) = p\mathcal{R}\tau[|A - V|^{p-1}U^*(-T)] \tag{3.2.3}$$

where $A - V = U|A - V|$ is the polar decomposition of $A - V$ (so that $\mathrm{Ker}U = \mathrm{Ker}|A - V|$ and U^*U is the orthogonal projection onto $(\mathrm{Ker}U)^\perp$). The arbitrary increment T of V is such that $A - (V + T) \in \mathcal{C}_p$ and $V + T$ is self-adjoint, that is, $T \in \mathcal{C}_p$ and T is self-adjoint. Take $T = s(e \otimes e)$ where $s \in \mathbb{R}$ and e is an arbitrary vector in H. Substituting into (3.2.3) and using (2.4.5) we get $0 = \mathcal{R}\langle|A - V|^{p-1}U^*e, e\rangle$. Hence, $\langle i|A - V|^{p-1}U^*e, e\rangle \in \mathbb{R}$, that is, $i|A - V|^{p-1}U^*$ is self-adjoint:

$$|A - V|^{p-1}U^* = -U|A - V|^{p-1}. \tag{3.2.4}$$

Now, $A + A^* - 2V = (A - V) + (A - V)^* = U|A - V| + |A - V|U^*$. Hence, to prove that $V = \mathcal{R}A$ we require to prove that

$$|A - V|U^* = -U|A - V| \tag{3.2.5}$$

which is what (3.2.4) says in the case $p = 2$. Otherwise, we appeal to the functional calculus described in Sect. 2.2. Write $Z = |A - V|^{p-1}$. Then (3.2.4) says that

$$ZU^* = -UZ \tag{3.2.6}$$

and we require to obtain (3.2.5), namely

$$Z^{\frac{1}{p-1}}U^* = -UZ^{\frac{1}{p-1}}. \tag{3.2.7}$$

This will follow by the functional calculus from

$$Z^nU^* = -UZ^n, \qquad n \in \mathbb{N}; \tag{3.2.8}$$

for the function $f : t \to t^{\frac{1}{p-1}}$, where $1 < p < \infty$ and $t \in \mathbb{R}^+ \supseteq \sigma(Z)$, can be approximated by a sequence $\{f_n\}$ of polynomials without constant term (for $f(0) = 0$); and if (3.2.8) holds then $f_n(Z)U^* = -Uf_n(Z)$ and hence, by (2.2.3), equality (3.2.7) holds.

We first prove (3.2.8) for odd n and then use the functional calculus to deduce (3.2.8) for even n. So we first show that

$$Z^{2k-1}U^* = -UZ^{2k-1}, \qquad k \in \mathbb{N}. \tag{3.2.9}$$

Assertion: $Z^2 U = UZ^2$. To prove this, note that in the polar decomposition of $A - V = U|A - V|$ we have Ker $U = $ Ker $|A - V| = $ Ker $Z^{\frac{1}{p-1}} = $ Ker Z (Recall that for a positive operator Z we have, by (2.2.1), Ker $Z^r = $ Ker Z for $0 < r < \infty$). Hence, $(\text{Ker } U)^\perp = (\text{Ker } Z)^\perp = $ Ran Z. Thus, U^*U, the orthogonal projection onto $(\text{Ker } U)^\perp$, satisfies $U^*UZ = Z = ZU^*U$. Hence, on multiplying (3.2.6) on the right by U and on the left by $-U^*$ we get $-U^*ZU^*U = U^*UZU$, that is, $-U^*Z = ZU$. Then, using (3.2.6) in the form $-ZU^* = UZ$, we get

$$Z^2 U = Z(ZU) = Z(-U^*Z) = -ZU^*Z = (UZ)Z = UZ^2.$$

The proof of (3.2.9) itself is now a simple induction: for $k = 1$, (3.2.9) is the same as (3.2.6); whilst if (3.2.9) holds for $k = r$ then it holds for $k = r + 1$ since, by the assertion,

$$Z^{2(r+1)-1}U^* = Z^2 Z^{2r-1}U^* = Z^2(-UZ^{2r-1}) = -UZ^2 Z^{2r-1} = -UZ^{2(r+1)-1}.$$

Finally, to deduce from (3.2.9) that (3.2.8) holds for even n we rewrite (3.2.9) in the form

$$(Z^{2q})(Z^{2k-1}U^*) = (-UZ^{2k-1})(Z^{2q})$$

for $q = 0, 1, 2, \ldots$ and for all k in $\mathbb{N}$. Hence, for every polynomial p

$$p(Z^2)(Z^{2k-1}U^*) = (-UZ^{2k-1})p(Z^2).$$

In particular, this holds for every member of a sequence $\{p_n\}$ of polynomials converging to the square root function $t \to \sqrt{t}$. Hence, $ZZ^{2k-1}U^* = -UZ^{2k-1}Z$, that is, $Z^{2k}U^* = -UZ^{2k}$, that is, (3.2.8) holds for even n. $\qquad\square$

In the case $p = 1$ provided the (somewhat restrictive) invertibility and finite-dimensionality conditions of Theorem 2.4.4 are met it follows from Theorem 3.2.4 that $X = \mathcal{R}A$ is **a** critical point of $X \mapsto \|A - X\|_1$; but the proof of Theorem 3.2.4 showing conversely that if V is a critical point then $V = \mathcal{R}A$ does not hold for $p = 1$ since the functional calculus argument explicitly requires that $p \neq 1$. For instance, in Example 3.2.2, the map $X \mapsto \|A - X\|$, where $A = \begin{bmatrix} 1 & 1 \\ 0 & 1 \end{bmatrix}$, is differentiable at one of its global minima, viz $X = \mathcal{R}A$ (since $A - \mathcal{R}A$ is invertible), but is not differentiable at the other global minimizer, viz $X = I$ (since $A - I$ is not invertible).

3.3   Positive Approximants

Positive operator approximation is more complicated than is self-adjoint approximation. To motivate Halmos' striking result [21, Theorem 1] (Theorem 3.3.1 below)
it is necessary to consider complex-valued functions. Let α be a given, bounded
complex-valued function

$$\alpha \colon \mathbb{R} \to \mathbb{C};$$

then a positive approximant of α is a positive function π which minimizes the radius
of the smallest closed disc, centre $(0,0)$, that includes the range of $\alpha - \pi$. Now,
to subtract a positive number from a complex number α, say, means to push α
horizontally to the left. Hence, if $\alpha(x) = \beta(x) + i\gamma(x)$, with $\beta(x)$ and $\gamma(x)$ real,
then the range of $\alpha - \pi$ cannot be covered by a disc with centre $(0,0)$ and smaller
radius than

$$r^- = \sup\{|\alpha(x)| : \beta(x) \le 0\}$$

nor by a disc with centre $(0,0)$ and smaller radius than

$$r^+ = \sup\{|\gamma(x)| : \beta(x) \ge 0\}.$$

Let $r = \max\{r^-, r^+\}$ and $\pi = \beta + (r^2 - \gamma^2)^{\frac{1}{2}}$; then $\|\alpha - \pi\| = r$ (For a bounded,
complex-valued function f, $\|f\| = \sup |f(x)|$). Thus, π is a positive approximant of
α, indeed, the largest one: for all positive functions p

$$\|\alpha - (\beta + (r^2 - \gamma^2)^{\frac{1}{2}})\| \le \|\alpha - p\|.$$

Theorem 3.3.1 *For $A = B + iC$ in $\mathcal{L}(H)$ let*

$$P(r) = B + (r^2 I - C^2)^{\frac{1}{2}},$$
$$\mathcal{N}(A) = \inf\{r : r \ge 0 \text{ and } P(r) \ge 0\}.$$

Then for all positive P in $\mathcal{L}(H)$

$$\|A - P(\mathcal{N}(A))\| \le \|A - P\|.$$

Proof Let

$$\delta(A) = \inf\{\|A - P\| : P \ge 0\}$$

First, it is not obvious that positive approximants always exist, i.e. that the
infimum defining $\delta(A)$ is always attained. We prove existence as follows. Observe

that (i) since $\delta(A) \leq \|A - 0\|$ there is no loss of generality in restricting attention to positive operators in the balls with centre A and radius $\|A\|$; (ii) the positive operators form a weakly closed set; and (iii) the norm $\|A - P\|$ is a weakly, lower semi-continuous function of P; from (i), (ii) and (iii) and the weak compactness of closed balls it follows that $\|A - P\|$ attains its infimum.

To prove the actual result of the theorem we prove that $\delta(A) = \mathcal{N}(A)$.

$\delta(A) \leq \mathcal{N}(A)$

The positive numbers of which $\mathcal{N}(A)$ is an infimum must satisfy two conditions: (i) $r^2 I - C^2 \geq 0$ (so that the positive square root $(r^2 I - C^2)^{\frac{1}{2}}$ exists); and (ii) $B + (r^2 I - C^2)^{\frac{1}{2}} \geq 0$. Positive numbers satisfying these two conditions always exist. For if $r^2 \geq \|B\|^2 + \|C\|^2$ then $r^2 I - C^2 \geq r^2 - \|C\|^2 \geq \|B\|^2 (\geq 0)$ and hence

$$B + (r^2 I - C^2)^{\frac{1}{2}} \geq B + \|B\| \geq 0$$

(we use here, and elsewhere, the inequality $\|X\| \geq \pm X$ valid for self-adjoint X).

Accordingly, let r be a positive number such that $r^2 I - C^2 \geq 0$ and such that if $P = B + (r^2 I - C^2)^{\frac{1}{2}}$ then $P \geq 0$. Since in $A - P = -(r^2 I - C^2)^{\frac{1}{2}} + iC$ the operators $-(r^2 I - C^2)^{\frac{1}{2}}$ and iC commute it follows that

$$\|A - P\|^2 = \|(A - P)^*(A - P)\| = \|(r^2 I - C^2) + C^2\| = r^2.$$

By definition, $\delta(A) \leq \|A - P\|$ for all positive P. Therefore, $\delta(A) \leq r$ and hence, taking the infimum over r, $\delta(A) \leq \mathcal{N}(A)$.

$\mathcal{N}(A) \leq \delta(A)$

As for the opposite inequality, let P be a positive operator. Since P is self-adjoint and $B(= \mathcal{R}A)$ is a self-adjoint approximant of A (by Theorem 3.2.1) it follows that $\|A - P\| \geq \|A - B\| = \|C\| \geq C$. Therefore, if $r = \|A - P\|$ then $r^2 I - C^2 \geq 0$, and hence $(r^2 I - C^2)^{\frac{1}{2}} \geq 0$. We show additionally that

$$B + (r^2 I - C^2)^{\frac{1}{2}} \geq 0; \tag{3.3.1}$$

for once (3.3.1) is proved then $\mathcal{N}(A) \leq r(= \|A - P\|)$ and hence, taking the infimum over r, $\mathcal{N}(A) \leq \delta(A)$.

To prove (3.3.1) we make use of the fact that if X and Y are self-adjoint then

$$\|X + iY\|^2 \geq \|X^2 + Y^2\| \tag{3.3.2}$$

(To see (3.3.2), let $Z = X + iY$ so that $\frac{1}{2}(Z^*Z + ZZ^*) = X^2 + Y^2$ and hence $\|X^2 + Y^2\| \leq \frac{1}{2}(\|Z^*Z\| + \|ZZ^*\|) = \|Z\|^2 = \|X + iY\|^2$.) Applying (3.3.2) to $A - P = (B - P) + iC$ we get $r^2 = \|A - P\|^2 \geq \|(B - P)^2 + C^2\| \geq (B - P)^2 + C^2$ and so

$$(r^2 I - C^2)^{\frac{1}{2}} \geq |B - P|$$

(the absolute value of a self-adjoint operator is the positive square root of its square). Decompose $B - P$ into positive and negative parts, $(B - P)^+$ and $(B - P)^-$, both of which are positive operators (cf. $f = f^+ - f^-$ and $|f| = f^+ + f^-$ for functions). Then

$$B + (r^2 I - C^2)^{\frac{1}{2}} \geq B + |B - P| = (B - P) + P + |B - P|$$
$$= (B - P)^+ - (B - P)^- + P + (B - P)^+ + (B - P)^-$$
$$= 2(B - P)^+ + P \geq 0$$

as desired (since $(B - P)^+$ and P are both positive operators).

Conclusion so far: $\mathcal{N}(A) = \delta(A)$

Finally, with $P(\mathcal{N}(A)) = B + (\mathcal{N}(A)^2 I - C^2)^{\frac{1}{2}}$ the operator $P(\mathcal{N}(A))$ is positive and since $\mathcal{N}(A) = \delta(A)$ we can find a sequence $\{r_n\}$ consisting of positive numbers and such that $\{r_n\} \searrow \delta(A)(= \mathcal{N}(A))$ and hence such that $B + (r_n^2 I - C^2)^{\frac{1}{2}} \searrow P(\mathcal{N}(A))$. $\square$

Observe that the proof of the Theorem 3.3.1 shows not only that the infima of the two sets

$$\{\|A - P\| : P \geq 0\} \quad \text{and} \quad \{r : r \geq 0 \text{ and } B + (r^2 I - C^2)^{\frac{1}{2}} \geq 0\}$$

are the same but that the sets, too, are the same. The point of Theorem 3.3.1 is that the infimum $\mathcal{N}(A)$, involving as it does positive numbers, is easier to calculate than the infimum $\delta(A)$, involving as it does positive operators (see Exercise 1).

Unfortunately, the result of Theorem 3.3.1 cannot be extended to $\mathcal{C}_p$. For with $P(r) = B + (r^2 I - C^2)^{\frac{1}{2}}$ then $|A - P(r)|^2 = r^2 I$ so that $A - P(r)$ is not compact. Even in finite dimensions the $\|\cdot\|_p$ analogue of Theorem 3.3.1, viz

$$\|A - P(\mathcal{N}(A))\|_p \leq \|A - P\|_p, \quad 1 \leq p < \infty$$

for $P \geq 0$, breaks down. For let $A = \begin{bmatrix} 0 & 0 \\ 1 & 0 \end{bmatrix}$; then $P(\mathcal{N}(A)) = \frac{1}{2}\begin{bmatrix} 1 & 1 \\ 1 & 1 \end{bmatrix}$ (see Exercise 1); and take $P = \frac{1}{4}\begin{bmatrix} 1 & 1 \\ 1 & 1 \end{bmatrix}$; then $P \geq 0$ and

$$\|A - P\|_2 = \sqrt{\frac{3}{4}} < 1 = \|A - P(\mathcal{N}(A))\|_2.$$

We can obtain a simpler result than Theorem 3.3.1 (Theorem 3.3.3 below) on positive approximation that does extend to $\mathcal{C}_p$ if we suppose that the operator to be approximated to is normal: Theorem 3.3.3 says that if A is normal then $(\mathcal{R}A)^+$ is a positive approximant of A. (Here, $(\mathcal{R}A)^+$ is the positive part of A: thus, $(\mathcal{R}A) = (\mathcal{R}A)^+ - (\mathcal{R}A)^-$ where $(\mathcal{R}A)^+ \geq 0$ and $(\mathcal{R}A)^- \geq 0$). Theorem 3.3.3 requires the following lemma.

Lemma 3.3.2 *Let X, Y and Z be self-adjoint with $0 \leq X \leq Y$. Then*

$$w(X + iZ) \leq w(Y + iZ)$$

(where w is the numerical radius); and if, further, $X + iZ$ is normal then

$$\|X + iZ\| \leq \|Y + iZ\|.$$

Proof Now, with w denoting the numerical radius (see last paragraph of Sect. 2.1)

$$|\langle (X + iZ)f, f \rangle|^2 = \langle Xf, f \rangle^2 + \langle Zf, f \rangle^2 \leq \langle Yf, f \rangle^2 + \langle Zf, f \rangle^2$$
$$= |\langle (Y + iZ)f, f \rangle|^2 \leq w(Y + iZ)^2$$

on taking $\|f\| = 1$. This yields the first assertion; and the second assertion follows immediately since $w(T) \leq \|T\|$ for all T with equality for normal T.   □

Theorem 3.3.3 ([21, Theorem 2])   *Let A be normal. Then for all positive P*

$$\|A - (\mathcal{R}A)^+\| \leq \|A - P\|.$$

Proof Let $B = \mathcal{R}A$ and $C = \mathcal{I}mA$. Decompose the underlying Hilbert space so that B may be written

$$B = \begin{bmatrix} B_1 & 0 \\ 0 & B_2 \end{bmatrix}$$

where B_1 and B_2 are positive and

$$(\mathcal{R}A)^+ = \begin{bmatrix} B_1 & 0 \\ 0 & 0 \end{bmatrix} \quad \text{and} \quad (\mathcal{R}A)^- = \begin{bmatrix} 0 & 0 \\ 0 & B_2 \end{bmatrix}.$$

As $A(= B + iC)$ is normal then $BC = CB$ and with respect to this decomposition

$$C = \begin{bmatrix} C_1 & 0 \\ 0 & C_2 \end{bmatrix}$$

where C_i are self-adjoint and $B_iC_i = C_iB_i$ for $i = 1, 2$. The arbitrary positive operator P may be written

$$P = \begin{bmatrix} P_1 & Q \\ Q^* & P_2 \end{bmatrix}$$

where P_1 and P_2 are positive. Thus,

$$\|A - P\| = \|(RA)^+ - (RA)^- - P\| = \left\| \begin{bmatrix} (B_1 - P_1) + iC_1 & -Q \\ -Q^* & (-B_2 - P_2) + iC_2 \end{bmatrix} \right\|$$

$$\geq \max\{\|(B_1 - P_1) + iC_1\|, \| - (B_2 + P_2) + iC_2\|\}.$$

As P_1 is self-adjoint it cannot be strictly closer to $B_1 + iC_1$ than B_1 (by Theorem 3.2.1 with $\|\|\cdot\|\| = \|\cdot\|$): that is $\|(B_1 - P_1) + iC_1\| \geq \|iC_1\|$. Applying Lemma 3.3.2 with "X" $= B_2$, "Y" $= B_2 + P_2$ and "Z" $= -iC_2$ yields

$$\| - (B_2 + P_2) + iC_2\| = \|Y + iZ\| \geq \|X + iZ\| = \| - B_2 + iC_2\|.$$

Thus:

$$\|A - P\| \geq \max\{\|iC_1\|, \| - B_2 + iC_2\|\}$$

$$= \left\| \begin{bmatrix} iC_1 & 0 \\ 0 & -B_2 + iC_2 \end{bmatrix} \right\| = \left\| \begin{bmatrix} (B_1 + iC_1) & 0 \\ 0 & -B_2 + iC_2 \end{bmatrix} - \begin{bmatrix} B_1 & 0 \\ 0 & 0 \end{bmatrix} \right\|$$

$$= \|A - (RA)^+\|.$$

$\square$

The extension of this result to C_p is an easy consequence of the theory of spectral approximants to be developed in Chap. 5 (this theory also yields a much shorter proof of Theorem 3.3.3 itself). For the moment the reader is invited (in Exercise 2) to derive—under slightly restrictive hypotheses—a partial extension to C_p of Theorem 3.3.3.

Exercises

1 For the matrix $A = \begin{bmatrix} 0 & 0 \\ 1 & 0 \end{bmatrix}$ prove that

(a) $\frac{1}{2}I$ is the smallest scalar that can be added to RA to produce a positive operator;
(b) $\delta(A) = \sqrt{2}/2$ (where $\delta(A)$ is as in Theorem 3.3.1);
(c) $P(N(A)) = \frac{1}{2}\begin{bmatrix} 1 & 1 \\ 1 & 1 \end{bmatrix}$ (where $N(A)$ is as in Theorem 3.3.1).

2 (a) Let X, Y and Z be self-adjoint and such that $0 \leq X \leq Y$, $X + iZ$ is normal and compact and $Y + iZ \in C_p$ where $1 \leq p < \infty$. Prove that $X + iZ \in C_p$ and

$$\|X + iZ\|_p \leq \|Y + iZ\|_p$$

with equality if, and for $1 < p < \infty$ only if, $Y = X$.

(b) Let A be normal, $(\mathcal{R}A)^+$ be compact and P be positive and such that $A - P \in \mathcal{C}_p$ where $1 \leq p < \infty$. Prove that $A - (\mathcal{R}A)^+ \in \mathcal{C}_p$ and that, provided $2 \leq p < \infty$,

$$\|A - (\mathcal{R}A)^+\|_p \leq \|A - P\|_p$$

with equality if and only if $P = (\mathcal{R}A)^+$.

Notes

The significance of self-adjoint approximation to quantum chemistry is explained in [18] (Section 1 is based on a personal communication from Professors Goldstein and Levy). The local result, Theorem 3.2.4, is reported in [30].

Halmos' deep result on positive approximation (Theorem 3.3.1) and its variant concerning positive approximation of normal operators (Theorem 3.3.3) are in [21, Theorems 1 and 2 respectively]. Theorem 3.3.3 has been generalized to approximation with respect to unitary invariant norms by Bhatia and Kittaneh [9, Theorem 2].

Exercise 1 is drawn from Halmos [21, p. 955].

Chapter 4
Commutator Approximants

We study approximation by commutators $AX - XA$, by generalized commutators $AX - XB$ and by self-commutators $X^*X - XX^*$ for varying X in the context of $\mathcal{L}(H)$ and $\mathcal{C}_p$.

These problems are precipitated by the commutation relation of quantum mechanics. This says that for certain linear **transformations** P and Q

$$PQ - QP = \alpha I$$

where α is a small, non-zero scalar. The Wielandt–Wintner theorem (Theorem 4.2.1) says that the equality $PQ - QP = \alpha I$ cannot hold for a non-zero scalar α and A and B in $\mathcal{L}(H)$ (i.e, for **bounded**, linear operators A and B). This suggests the problem of approximating the identity I by the commutator $AX - XA$ for A and (varying) X in $\mathcal{L}(H)$: the simplest commutator approximant problem (solved in Theorem 4.3.2 below).

4.1 The Commutation Relation of Quantum Mechanics and the Heisenberg Uncertainty Principle

In this section we motivate, in the simplest case possible, the commutation relation of quantum mechanics. Our treatment is non-relativistic. Probabilistic notions are involved in the setting up of the foundations of quantum mechanics: thus, quantum mechanics is non-deterministic, in contrast to Newtonian, and relativistic, mechanics, both of which are deterministic.

We outline, in a non-relativistic setting, the mechanics of a particle moving in one dimension. A **state** ψ of a particle constrained to one dimension is a Lebesgue square-summable, complex-valued function on $\mathbb{R}$, i.e, $\psi \in \mathcal{L}^2(-\infty, \infty)$. The space

© Springer International Publishing AG 2017
P.J. Maher, *Operator Approximant Problems Arising from Quantum Theory*,
DOI 10.1007/978-3-319-61170-9_4

$\mathcal{L}^2(-\infty, \infty)$ is sometimes called the corresponding **state space**. If S is a Lebesgue measurable subset of $\mathbb{R}$ then $\int_S |\psi(x)|^2 dx$ is the probability that a particle in state ψ is (physically) in the set S. Therefore, since the particle has a probability of 1 being **somewhere** in $\mathbb{R}$ we must have $\int_{-\infty}^{\infty} |\psi(x)|^2 dx = 1$. Thus:

$$\|\psi\|^2 = \langle \psi, \psi \rangle = \int_{-\infty}^{\infty} |\psi(x)|^2 dx = 1.$$

Given a random variable, that is, a quantity that takes on real-values in some probabilistic setting, its **probability density function** is a function $f : \mathbb{R} \to \mathbb{R}$ such that $\int_{x_1}^{x_2} f(x) dx$ is the probability that the quantity is between x_1 and x_2. Therefore, from the previous paragraph, the function $x \to |\psi(x)|^2$ is the probability density function corresponding to the random variable whose values are locations in $\mathbb{R}$ of a particle in state ψ (more concisely, we say $x \to |\psi(x)|^2$ is the probability density function of the position of the particle in state ψ).

If $f : \mathbb{R} \to \mathbb{R}$ is the probability density function arising from a random variable, the **expectation** E_f (sometimes called expected value E_f) is defined by

$$E_f = \int_{-\infty}^{\infty} xf(x)dx. \tag{4.1.1}$$

The **variance** D_f is defined by

$$D_f = \int_{-\infty}^{\infty} (x - E_f)^2 f(x) dx \tag{4.1.2}$$

and $(D_f)^{\frac{1}{2}}$ is called the **standard deviation** for the random variable (These definitions are the obvious analogues of the familiar notions of expectation, variance and standard deviation of ordinary statistics).

We introduce the concept of an observable (sometimes called a dynamical variable). Examples of observables are: position, momentum, energy. In the case of our particle constrained to move in one dimension, and its state space $\mathcal{L}^2(-\infty, \infty)$, think of an **observable** as a linear map $A : \mathcal{L}^2(-\infty, \infty) \to \mathcal{L}^2(-\infty, \infty)$ (we shall see that certain observable are not defined on all of $\mathcal{L}^2(-\infty, \infty)$ as bounded, linear operators; although, for the moment, we can treat them as such). We denote the expectation of an observable A, given the state ψ, by $E_\psi(A)$ and, by analogy with (4.1.1), define $E_\psi(A)$ by

$$E_\psi(A) = \langle A\psi, \psi \rangle .$$

Let Q denote the position observable of a particle in state ψ. Then

$$E_\psi(Q) = \langle Q\psi, \psi \rangle = \int_{-\infty}^{\infty} Q\psi(x)\overline{\psi(x)}dx. \tag{4.1.3}$$

Since $x \rightarrow |\psi(x)|^2$ is the probability density function of the position of the particle in state ψ, it follows from (4.1.1) that $E_{|\psi|^2}$ is the expectation of the position of the particle in state ψ so that $E_{|\psi|^2} = E_\psi(Q)$ and, by (4.1.1),

$$E_{|\psi|^2} = \int_{-\infty}^{\infty} x|\psi(x)|^2 dx = \int_{-\infty}^{\infty} x\psi(x)\overline{\psi(x)}dx. \qquad (4.1.4)$$

Comparing (4.1.3) and (4.1.4) suggests we define the position operator Q on $\mathcal{L}^2(-\infty, \infty)$ by

$$(Q\psi)(x) = x\psi(x).$$

Since $E_{|\psi|^2}(= E_\psi(Q))$ is real it follows from (4.1.3) that Q, where defined, is self-adjoint.

There is a more complicated physical argument, which we omit, that justifies defining the momentum operator P on $\mathcal{L}^2(-\infty, \infty)$ by

$$(P\psi)(x) = \frac{\hbar}{2\pi i} \frac{d\psi(x)}{dx}$$

where $\hbar = 6.665 \times 10^{-34}$ Joule/sec is Planck's constant. The operator P, like Q, is not defined everywhere on $\mathcal{L}^2(-\infty, \infty)$ but, where defined, is linear and can be shown to be self-adjoint.

We now obtain the commutation relation (4.1.5) of quantum mechanics. Let $\alpha = \frac{\hbar}{2\pi i}$ so that $(P\psi)(x) = \alpha \frac{d\psi(x)}{dx}$. Then, for all ψ in $\mathcal{L}^2(-\infty, \infty)$,

$$(PQ - QP)\psi(x) = P(Q\psi)(x) - Q(P\psi)(x)$$
$$= P(x\psi(x)) - Q(\alpha \frac{d\psi(x)}{dx})$$
$$= \alpha \frac{d(x\psi(x))}{dx} - \alpha x \frac{d\psi(x)}{dx}$$
$$= \alpha \left[x \frac{d\psi(x)}{dx} + 1.\psi(x) - x \frac{d\psi(x)}{dx} \right] = \alpha\psi(x).$$

Therefore,

$$PQ - QP = \alpha I \qquad (4.1.5)$$

where $\alpha = \frac{\hbar}{2\pi i}$, $\hbar$ being Planck's constant, and P is the momentum operator and Q is the position operator on $\mathcal{L}^2(-\infty, \infty)$.

From the commutation relation (4.1.5) we obtain, via Lemma 4.1.1, the celebrated Heisenberg uncertainty principle, Corollary 4.1.2 (the inclusion of the

Heisenberg uncertainty principle is a luxury in this book since nowhere else is it referred to). Lemma 4.1.1 is couched in terms of self-adjoint operators on a Hilbert space H. If A is self-adjoint we define its expectation $E_\psi(A)$ by $E_\psi(A) = \langle A\psi, \psi \rangle$, where $\psi \in H$, and, by analogy with (4.1.2), its variance $D_\psi(A)$ by

$$D_\psi(A) = \|(A - E_\psi(A)I)\psi\|^2$$

$$= \langle (A - E_\psi(A)I)\psi, (A - E_\psi(A)I)\psi \rangle$$

$$(= \int_{-\infty}^{\infty} (A - E_\psi(A)I)^2 \psi(x)\overline{\psi(x)}dx)$$

this last equality holding in the special case $H = \mathcal{L}^2(-\infty, \infty)$.

Lemma 4.1.1 *If A and B are self-adjoint then*

$$D_\psi(A)D_\psi(B) \geq \frac{1}{4}|E_\psi(AB - BA)|^2.$$

Proof Now, as A and B are self-adjoint,

$$|E_\psi(AB - BA)|^2 = |\langle (AB - BA)\psi, \psi \rangle|^2$$

$$= |\langle AB\psi, \psi \rangle - \langle \psi, AB\psi \rangle|^2$$

$$= |\langle AB\psi, \psi \rangle - \overline{\langle AB\psi, \psi \rangle}|^2$$

$$= (2\mathcal{I}\text{m} \langle AB\psi, \psi \rangle)^2. \qquad (4.1.6)$$

We now substitute for $AB - BA$ the following identity valid for arbitrary real a and b:

$$AB - BA = (A - aI)(B - bI) - (B - bI)(A - aI).$$

Then letting $a = E_\psi(A)$ and $b = E_\psi(B)$ we get

$$\frac{1}{4}|E_\psi(AB - BA)|^2 = \frac{1}{4}|E_\psi((A - E_\psi(A)I)(B - E_\psi(B)I)$$

$$- (B - E_\psi(B)I)(A - E_\psi(A)I))|^2$$

$$= (\mathcal{I}\text{m} \langle (A - E_\psi(A)I)(B - E_\psi(B)I)\psi, \psi \rangle)^2$$

$$\leq \|(A - E_\psi(A)I)\psi\|^2\|(B - E_\psi(B)I)\psi\|^2$$

$$= D_\psi(A)D_\psi(B)$$

where the second equality comes from (4.1.6) with "A" $= A - E_\psi(A)I$ and "B" $= B - E_\psi(B)I$ and the inequality comes from the Cauchy–Schwartz inequality. $\qquad \square$

Corollary 4.1.2 (Heisenberg Uncertainty Principle) *Let P be the momentum observable and Q be the position observable of a particle moving in one dimension. Then*

$$D_\psi^{\frac{1}{2}}(P)D_\psi^{\frac{1}{2}}(Q) \geq \frac{\hbar}{4\pi}$$

where $\hbar$ is Planck's constant.

Proof With Q the position observable and P the momentum observable then, by (4.1.5), $PQ - QP = \dfrac{\hbar I}{2\pi i}$ so that, from Lemma 4.1.1,

$$D_\psi^{\frac{1}{2}}(P)D_\psi^{\frac{1}{2}}(Q) \geq \frac{1}{2}\left|E_\psi(\frac{\hbar I}{2\pi i})\right| = \frac{1}{2}\left|\left\langle \frac{\hbar I}{2\pi i}\psi, \psi\right\rangle\right| = \frac{\hbar}{4\pi}\|\psi\|^2 = \frac{\hbar}{4\pi}.$$

$\square$

Physicists sometimes write Δp and Δq for $D_\psi^{\frac{1}{2}}(P)$ and $D_\psi^{\frac{1}{2}}(Q)$. The Heisenberg uncertainty principle then says that $(\Delta p)(\Delta q) \geq \frac{\hbar}{4\pi}$: thus, the momentum and position of a particle cannot be determined simultaneously with complete accuracy.

As commented earlier, the position observable Q and the momentum observable P are not bounded linear operators on $\mathcal{L}^2(-\infty, \infty)$. The derivation of Corollary 4.1.2 is, however, valid in the sense that Lemma 4.1.1 is valid for vectors at which the operators in question are defined.

4.2 Wielandt–Wintner Theorem

Consider the equality

$$AB - BA = \alpha I \tag{4.2.1}$$

of Sect. 4.1. Recalling the derivation of it in Sect. 4.1, it holds if the underlying space $H = \mathcal{L}^2(-\infty, \infty)$ and if A and B are the (unbounded) operators given by $(Af)(x) = f'(x)$ and $(Bf)(x) = xf(x)$ where $f \in \mathcal{L}^2(-\infty, \infty)$.

But (4.2.1) cannot be satisfied for non-zero α if A and B are finite square matrices. To see this, take the trace τ of both sides of (4.2.1); then, because trace is linear, multiplicative and $\tau(AB) = \tau(BA)$, we have $\tau(AB - BA) = 0$ and $\tau(\alpha I) \neq 0$: contradiction.

For some years in the 1930s and 1940s much interest centred on whether not (4.2.1) could hold for bounded, linear operators A and B. This was settled, independently, by Wielandt [44], in 1949, and by Wintner [45], in 1947 (Significantly, Wintner's paper is in a leading *Physics* journal). Both proofs are beautiful but we give only Wielandt's: its arguments are more elementary than those of Wintner and, as we shall see, permeate much of the work on commutator approximation.

Theorem 4.2.1 (Wielandt–Wintner)   *Let A and B be in $\mathcal{L}(H)$. If $AB - BA = \alpha I$ then $\alpha = 0$.*

Proof The trick is to form powers $A^n B - BA^n$. If $AB - BA = \alpha I$ then

$$A^2 B - BA^2 = A(AB - BA) + (AB - BA)A = 2\alpha A$$

and therefore, by induction,

$$A^n B - BA^n = n\alpha A^{n-1} \qquad \text{if} n \in \mathbb{N}; \tag{4.2.2}$$

for if $A^k B - BA^k = k\alpha A^{k-1}$ for $k = 1, \ldots, n$ then

$$A^{k+1} B - BA^{k+1} = A(A^k B - BA^k) + (AB - BA)A^k = \alpha(k + 1)A^k.$$

In the special case when A is nilpotent of index n (meaning that $A^n = 0$ and $A^{n-1} \neq 0$) then (4.2.2) forces $n\alpha A^{n-1} = 0$ so that $\alpha = 0$. If A is not nilpotent then, on taking norms, (4.2.2) gives inequality

$$n|\alpha| \|A^{n-1}\| \leq 2\|A^n\| \|B\| \leq 2\|A^{n-1}\| \|A\| \|B\|$$

so that, on dividing through by $n\|A^{n-1}\|$, we get

$$|\alpha| \leq 2\frac{\|A\| \|B\|}{n}.$$

As this holds for all $n = 1, 2, \ldots$ this forces $\alpha = 0$. $\square$

It can be checked that the above proof holds for an arbitrary complex normed algebra with unit. In particular, Theorem 4.2.1 applies to the Calkin algebra $\mathcal{K}(H)$ (the normed algebra $\mathcal{L}(H)$ factored by the ideal of compact operators).

Corollary 4.2.2 *For A and B in $\mathcal{L}(H)$ and for a non-zero scalar α, the operator $\alpha I - (AB - BA)$ is not compact.*

Proof In the Calkin algebra $\mathcal{K}(H)$ a non-zero scalar α is not a commutator by Theorem 4.2.1. Translated back to bounded operators this means that the difference between a non-zero scalar and a commutator is not compact. $\square$

A feature of Wielandt's proof of Theorem 4.2.1 is that the commutator $AB-BA(=\alpha I$ in the hypothesis of Theorem 4.2.1) commutes with A (and B). A generalization is to consider a commutator $AB - BA$ that commutes with one of the operators, A say. In [23, Problem 232] Halmos shows how this leads to the Kleinecke–Shirokov Theorem [26, 42] which we now state. (Recall: an operator A is quasinilpotent if $\lim_{n \to \infty} \|A^n\|^{\frac{1}{n}} = 0$, that is, if $\sigma(A) = \{0\}$.)

Theorem 4.2.3 (Kleinecke–Shirokov)   *If $AB-BA$ commutes with A then $AB-BA$ is quasinilpotent.*

If, in addition to the hypotheses of Theorem 4.2.3, the operator A is normal then the commutator $AB-BA$ turns out to be the simplest quasinilpotent operator, the zero operator. Although Theorem 4.2.4 is a variant of the Kleinecke–Shirokov Theorem its method of proof is quite different.

Theorem 4.2.4 *If $AB - BA$ commutes with A and if A is normal then $AB - BA = 0$.*

Proof The proof hinges on the spectral resolution of the normal operator A. This says that there exists a spectral measure $E(\cdot)$ such that for each $\epsilon > 0$ there exists disjoint Borel sets S_i, where $1 \le i \le n$, such that

$$\text{if } \lambda_i \in S_i \text{ and } S = \sum_{i=1}^{n} \lambda_i E(S_i) \text{ then } \|A - S\| < \epsilon. \tag{4.2.3}$$

The operator $AB - BA$ commutes with A and hence with each of the spectral projections $E(S_i)$ and so, since $\sum_{1}^{n} E(S_i) = E(\cup S_i) = I$,

$$AB - BA = (AB - BA) \sum_{i=1}^{n} E(S_i) = \sum_{i=1}^{n} E(S_i)(AB - BA)E(S_i). \tag{4.2.4}$$

Since the Borel sets are pairwise disjoint, $E(S_i)E(S_j)$ equals $E(S_j)$ if $i = j$ and zero if $i \ne j$. Hence, on substituting for the sum S we find that for each fixed i

$$E(S_i)(SB - BS)E(S_i) = \lambda_i E(S_i)BE(S_i) - \lambda_i E(S_i)BE(S_i) = 0.$$

Hence,

$$\|E(S_i)(AB - BA)E(S_i)\| = \|E(S_i)[(SB - BS) + (A - S)B - B(A - S)]E(S_i)\|$$
$$\le 2\|A - S\|\|B\|\|E(S_i)\| < 2\epsilon\|B\|$$

from (4.2.3). Since, from (4.2.4), $\|AB-BA\| = \sup \|E(S_i)(AB-BA)E(S_i)\|$ therefore $AB - BA = 0$. $\square$

4.3 Commutator Approximants in $\mathcal{L}(H)$

Wielandt–Wintner (Theorem 4.2.1) says that a commutator of bounded operators cannot be equal to the identity. How close can such a commutator be to the identity? Interpreting "close" in terms of the supremum norm, we wish to minimize

$$\|I - (AX - XA)\|$$

(the notation "X", in place of the "B" of Sect. 4.2, is meant to indicate the dependence of the quantity $\|I - (AX - XA)\|$ on the variable operator X for fixed A). It is not hard to show that, in finite dimensions, $\|I - (AX - XA)\| \geq \|I\|$ (See Exercise 1). The same holds in infinite dimensions provided A commutes with $AX - XA$ or if, instead, A is hyponormal; see Theorem 4.3.2.

An operator A is said to be **hyponormal** if $A^*A \geq AA^*$, equivalently, if $\|A^*g\| \leq \|Ag\|$ for all g in H [23, Problem 203]. The following result—the power norm equality—is well known [23, Problem 205]; but we include a proof since Exercise 6(a) asks the reader to extend it to a wider class of operators.

Lemma 4.3.1 *If A is hyponormal then $\|A^n\| = \|A\|^n$ for each n in $\mathbb{N}$.*

Proof Equality is trivial for $n = 1$. Proceed by induction. Now

$$
\begin{aligned}
\|A^n f\|^2 = \langle A^n f, A^n f \rangle &= \langle A^* A^n f, A^{n-1} f \rangle \\
&\leq \|A^* A^n f\| \|A^{n-1} f\| \\
&\leq \|A^{n+1} f\| \|A^{n-1} f\| \qquad \text{[hyponormality of } A] \\
&\leq \|A^{n+1}\| \|A^{n-1}\| \|f\|
\end{aligned}
$$

for every f in H so that

$$
\|A^n\|^2 \leq \|A^{n+1}\| \|A^{n-1}\|.
$$

But by the induction hypothesis ($\|A^k\| = \|A\|^k$ for $1 \leq k \leq n$) this inequality is equivalent to

$$
\|A\|^{2n} \leq \|A^{n+1}\| \|A\|^{n-1}.
$$

whence $\|A\|^{n+1} \leq \|A^{n+1}\|$. Since the reverse inequality is automatic, the inductive step follows.   □

Theorem 4.3.2, which is due to Halmos [19, Theorems 6 and 7] is probably the earliest result on commutator approximation.

Theorem 4.3.2

(a) If A commutes with $AX - XA$ then

$$
\|I - (AX - XA)\| \geq \|I\|;
$$

(b) the same conclusion as in (a) holds if, instead, A is hyponormal.

Proof

(a) Suppose, on the contrary, that $\|I - (AX - XA)\| < 1 (= \|I\|)$. Then $(AX - XA)$ is invertible. Since A commutes with $(AX - XA)$ it commutes with $(AX - XA)^{-1}$. Thus:

$$1 = (AX - XA)(AX - XA)^{-1}$$
$$= A[X(AX - XA)^{-1}] - [X(AX - XA)^{-1}]A$$

contrary to Wielandt–Wintner.

(b) It can be checked that

$$A^n X - XA^n = \sum_{i=0}^{n-1} A^{n-i-1}(AX - XA)A^i$$

$$= nA^{n-1} - \sum_{i=0}^{n-1} A^{n-i-1}(I - (AX - XA))A^i$$

so that

$$nA^{n-1} = (A^n X - XA^n) + \sum_{i=0}^{n-1} A^{n-i-1}(I - (AX - XA))A^i.$$

Hence,

$$n\|A^{n-1}\| \leq 2\|A^{n-1}\|\|A\|\|X\| + \|I - (AX - XA)\| \sum_{i=0}^{n-1} \|A^{n-i-1}\|\|A^i\|. \qquad (4.3.1)$$

But since A is hyponormal then, by Lemma 4.3.1, the summation in (4.3.1) equals $n\|A^{n-1}\|$. Divide both sides of the inequality (4.3.1) by $n\|A^{n-1}\|$: then

$$1 \leq \frac{2\|A\|\|X\|}{n} + \|I - (AX - XA)\|.$$

Since this last inequality has to hold for all n it follows that $\|I - (AX - XA)\| \geq \|I\|$. $\qquad \square$

In the same way as in Theorem 4.3.2(b) we can minimize $\|T - (AX - XA)\|$ if A is isometric and commutes with T. It is convenient to use the following identity: if $AT = TA$ then

$$nA^{n-1}T = A^n B - BA^n + \sum_{i=0}^{n-1} A^{n-i-1}(T - (AB - BA))A^i \qquad (4.3.2)$$

for all B in $\mathcal{L}(H)$.

Theorem 4.3.3 *If A is an isometry and if $AT = TA$ then*

$$\|T - (AX - XA)\| \geq \|T\|.$$

Proof In (4.3.2) let $B = X$ and take norms of (4.3.2). Then, cf. proof of Theorem 4.3.2(b), since $\|A^k\| = 1$ (because A is isometric) it follows that $\sum_{i=0}^{n-1} \|A^{n-i-1}\| \|A^i\| = n$. Therefore, on dividing both sides by n, we get

$$\|T\| = \|A^{n-1}T\| \leq \frac{2\|X\|}{n} + \|T - (AX - XA)\|$$

Hence the result.   □

Theorem 4.3.3 (and Theorems 4.3.5 and 4.3.7 below) can be expressed geometrically. Let $\mathcal{M}$ and $\mathcal{N}$ be linear subsets of $\mathcal{L}(H)$. We say that $\mathcal{M}$ is **orthogonal** to $\mathcal{N}$, denoted $\mathcal{M} \perp \mathcal{N}$ if, for all m in $\mathcal{M}$ and for all n in $\mathcal{N}$

$$\|m + n\| \geq \|n\|.$$

For historical comments see [13, p. 93].

For fixed A in $\mathcal{L}(H)$ consider the map $\Delta_A : \mathcal{L}(H) \to \mathcal{L}(H)$ given by

$$\Delta_A(X) = AX - XA$$

for varying X in $\mathcal{L}(H)$. It is trivial that Δ_A is linear, continuous and that $\Delta_A(XY) = (\Delta_A(X))Y + X(\Delta_A(Y))$ (for this reason Δ_A is sometimes called a derivation). Let the linear subsets $\text{Ran}\Delta_A$ and $\text{Ker}\Delta_A$ be defined by

$$\text{Ran}\Delta_A = \{Y \in \mathcal{L}(H) : Y = \Delta_A(X) \text{ for varying } X \text{ in } \mathcal{L}(H)\},$$

$$\text{Ker}\Delta_A = \{X \in \mathcal{L}(H) : \Delta_A(X) = 0\}$$

(that is, $\text{Ker}\Delta_A$ is the commutant of A). The subset $\text{Ran}\Delta_A$ is not necessarily closed; $\text{Ker}\Delta_A$ is closed. Geometrically, Theorem 4.3.3 is as follows.

Corollary 4.3.4 *If A is an isometry and $\{0\} \neq \text{Ker}\Delta_A$ then $\text{Ran}\Delta_A \perp \text{Ker}\Delta_A$.*

We aim to extend Theorem 4.3.3 to minimizing the quantity $\|T - (AX - XA)\|$ for normal A commuting with T. The next step in this direction is Theorem 4.3.5. It is interesting that in the actual *proofs* of Theorems 4.3.5 and 4.3.7 below geometric reasoning is used.

Theorem 4.3.5 *Let A be self-adjoint and $AT = TA$. Then*

$$\|T - (AX - XA)\| \geq \|T\|.$$

Proof The proof uses the theory of the Cayley transform [3, p. 42, 94]. Let U be the (isometric) Cayley transform of the self-adjoint operator A given by $U = (A - i)(A + i)^{-1}$, equivalently, $A = i(I + U)(I - U)^{-1}$. Then, as $\Delta_A(X) = \Delta_{A+k}(X)$ for a constant k and as $A - i = U(A + i)$ we have

$$\Delta_A(X) = \Delta_{A-i}(X) = U(A + i)X - XU(A + i)$$
$$= U(A + i)X - (A + i)XU + (A + i)XU - XU(A + i)$$
$$= \Delta_U((A + i)X) + \Delta_{A+i}(XU)$$

so that

$$\Delta_U((A + i)X) = \Delta_A(X) - \Delta_{A+i}(XU)$$
$$= \Delta_{A+i}(X) - \Delta_{A+i}(XU)$$
$$= \Delta_{A+i}(X(1 - U))$$

by the linearity of Δ. Since $1 - U$ and $A + i$ are both invertible, $\mathrm{Ran}\Delta_A = \mathrm{Ran}\Delta_U$. Also, $AT = TA$ implies $UT = TU$ so that $(T \in) \mathrm{Ker}\Delta_A \subseteq \mathrm{ker}\Delta_U$. By Corollary 4.3.4, $\mathrm{Ran}\Delta_U \perp \mathrm{Ker}\Delta_U$ ($\supseteq \mathrm{Ker}\Delta_A$) so that $\mathrm{Ran}\Delta_A(= \mathrm{Ran}\Delta_U) \perp \mathrm{Ker}\Delta_A$, that is,

$$\|T - (AX - XA)\| \geq \|T\|.$$

$\square$

Lemma 4.3.6 *Let $P_1, \ldots, P_n$ be orthogonal idempotents (i.e. $P_iP_j = 0$ if $i \neq j$ and $P_i^2 = P_i$ for $i = 1, \ldots, n$), let $\{\lambda_1, \ldots \lambda_n\}$ and $\{\mu_1, \ldots, \mu_n\}$ be sets of non-zero complex numbers such that $\lambda_i \neq \lambda_j$ and $\mu_i \neq \mu_j$ if $i \neq j$ and let*

$$Q_1 = \sum_{i=1}^{n} \lambda_i P_i \quad and \quad Q_2 = \sum_{i=1}^{n} \mu_i P_i.$$

Then $\mathrm{Ran}\Delta_{Q_1} = \mathrm{Ran}\Delta_{Q_2}$.

Proof Let $P_0 = 1 - \sum_{i=1}^{n} P_i$ and $\lambda_0 = \mu_0 = 0$. Then, for arbitrary X in $\mathcal{L}(H)$, a computation verifies that

$$\Delta_{Q_1}(X) = \sum_{i=0}^{i=n} \sum_{j=0}^{j=n} (\lambda_i - \lambda_j) P_i X P_j$$

$$\Delta_{Q_2}(X) = \sum_{i=0}^{i=n} \sum_{j=0}^{j=n} (\mu_i - \mu_j) P_i X P_j$$

and since $\lambda_i \neq \lambda_j$ and $\mu_i \neq \mu_j$ for $i \neq j$ the result follows.

$\square$

Theorem 4.3.7 *Let A be normal and $AT = TA$. Then*

$$\|T - (AX - XA)\| \geq \|T\|.$$

Proof Let $E(\cdot)$ be the spectral measure of the normal operator A (so that for each $\epsilon > 0$ there exist disjoint Borel sets S_i for $1 \leq i \leq n$, such that if $\lambda_i \in S_i$ then $\|A - \sum_1^n \lambda_i E(S_i)\| < \epsilon$, (cf. proof of Theorem 4.2.4).) Then the result will be proved if

$$\left\| T - \left[\left(\sum_{i=1}^n \lambda_i E(S_i) \right) X - X \left(\sum_{i=1}^n \lambda_i E(S_i) \right) \right] \right\| \geq \|T\|$$

for all X in $\mathcal{L}(H)$, for all T in $\mathrm{Ker}\Delta_A$, for every collection $\{S_i\}$ of Borel sets and for every collection $\{\lambda_i\}$ of complex numbers. Let

$$Q_1 = \sum_{i=1}^n \lambda_i E(S_i).$$

Thus, the result will be proved if $\|T - \Delta_{Q_1}(X)\| \geq \|T\|$, that is, in geometric language, if $\mathrm{Ran}\Delta_{Q_1} \perp \mathrm{Ker}\Delta_A$.

Introduce the self-adjoint operator

$$Q_2 = \sum_{i=1}^n i E(S_i)$$

(where the i above is the real subscript). Since T commutes with A it commutes with each of the spectral projections $E(S_i)$ and hence with Q_1 and Q_2. As $TQ_2 = Q_2 T$ and as Q_2 is self-adjoint then, by Theorem 4.3.5, $\|T - (Q_2 X - XQ_2)\| \geq \|T\|$ that is, $\mathrm{Ran}\Delta_{Q_2} \perp \mathrm{Ker}\Delta_A$. But, on applying Lemma 4.3.6, $\mathrm{Ran}\Delta_{Q_1} = \mathrm{Ran}\Delta_{Q_2}$. Therefore, $\mathrm{Ran}\Delta_{Q_1} \perp \mathrm{Ker}\Delta_A$, as desired.   □

Corollary 4.3.8 *If A is normal and $\{0\} \neq \mathrm{Ker}\Delta_A$ then $\mathrm{Ran}\Delta_A \perp \mathrm{Ker}\Delta_A$.*

We now extend Theorem 4.3.7 to approximation by operators of the form $AX - XB$ (sometimes called generalized commutators). The result is due to Anderson and Foias [6, Theorem 1.5]: the proof we give is an adaptation of an argument of Bouali and Cherki [10, Lemma 2.1].

Theorem 4.3.9 *Let A and B be normal and such that $AT = TB$. Then*

$$\|T - (AX - XB)\| \geq \|T\|.$$

Proof We reduce the unknown (approximating by a generalized commutator) to the known (approximating by a commutator). On $H \oplus H$, let $L = \begin{bmatrix} A & 0 \\ 0 & B \end{bmatrix}$, $S = \begin{bmatrix} 0 & T \\ 0 & 0 \end{bmatrix}$ and $Y = \begin{bmatrix} 0 & X \\ 0 & 0 \end{bmatrix}$. Then L is normal and since $AT = TB$ then $LS = SL$. Hence, by Theorem 4.3.7, $\|S - (LY - YL)\| \geq \|S\|$, that is

$$\left\| \begin{bmatrix} 0 & T - (AX - XB) \\ 0 & 0 \end{bmatrix} \right\| \geq \left\| \begin{bmatrix} 0 & T \\ 0 & 0 \end{bmatrix} \right\|$$

whence $\|T - (AX - XB)\| \geq \|T\|$.   □

To interpret this geometrically in Corollary 4.3.10, for fixed A, B in $\mathcal{L}(H)$ define the linear map $\Delta_{A,B} : \mathcal{L}(H) \to \mathcal{L}(H)$ by

$$\Delta_{A,B}(X) = AX - XB$$

for varying X in $\mathcal{L}(H)$. Let the linear subsets $\mathrm{Ran}\Delta_{A,B}$ and $\mathrm{Ker}\Delta_{A,B}$ be defined by

$$\mathrm{Ran}\Delta_{A,B} = \{Y \in \mathcal{L}(H) : Y = \Delta_{A,B}(X) \text{ for   varying } X \text{ in } \mathcal{L}(H)\},$$

$$\mathrm{Ker}\Delta_{A,B} = \{X \in \mathcal{L}(H) : \Delta_{A,B}(X) = 0\}.$$

Corollary 4.3.10 *If A and B are normal and $\{0\} \neq \mathrm{Ker}\Delta_{A,B}$ then $\mathrm{Ran}\Delta_{A,B} \perp \mathrm{Ker}\Delta_{A,B}$.*

We now generalize Theorem 4.3.9 to Fuglede–Putnam pairs of operators. Recall from Chap. 2 Fuglede's Theorem: this says that if the operator T commutes with the normal operator A then T commutes with A^*. From this one can deduce Putnam's corollary, which we state below [23, Problem 192].

Corollary 4.3.11 *If A and B are normal operators and if T is such that $AT = TB$ then $A^*T = TB^*$.*

This suggests we generalize beyond normal operators by naming the concept embodied in the conclusion of Corollary 4.3.11. For generality we frame this definition in terms of a two-sided ideal $\mathcal{I}$ of $\mathcal{L}(H)$.

Definition 4.3.12 Let A and B be in $\mathcal{L}(H)$ and $\mathcal{I}$ be a two-sided ideal of $\mathcal{L}(H)$. The pair (A, B) is said to possess the **Fuglede–Putnam property**, denoted $(F\text{-}P)_{\mathcal{I}}$, if $AT = TB$ and $T \in \mathcal{I}$ implies $A^*T = TB^*$ (if $\mathcal{I} = \mathcal{L}(H)$ then the Fuglede–Putnam property is denoted $(F\text{-}P)$).

Geometrically: the pair has the property $(F\text{-}P)_{\mathcal{I}}$ if $\{0\} \neq \mathrm{Ker}\Delta_{A,B} \cap \mathcal{I}$ implies $(\mathrm{Ker}\Delta_{A,B} \cap \mathcal{I}) \subseteq \mathrm{Ker}\Delta_{A^*,B^*}$.

Obviously, if A and B are normal operators then, by Corollary 4.3.11, (A, B) possesses the Fuglede–Putnam property $(F\text{-}P)$. Rather deeper, is the following geometric result, Lemma 4.3.13, due to Bouali and cited in [10, Theorem 1.2].

Lemma 4.3.13 *Let A and B in $\mathcal{L}(H)$ and $\mathcal{I}$ be a two-sided ideal of $\mathcal{L}(H)$. Then the following statements, (a) and (b), are equivalent:*

(a) *(A, B) has the property $(F\text{-}P)_{\mathcal{I}}$;*
(b) *if $AT = TB$ and $T \in \mathcal{I}$ then $\overline{\mathrm{Ran}T}$ reduces A, $(\mathrm{Ker}T)^{\perp}$ reduces B and restrictions $A|_{\overline{\mathrm{Ran}T}}$ and $B|_{(\mathrm{Ker}T)^{\perp}}$ are normal operators.*

Theorem 4.3.14 *Let the pair (A, B) in $\mathcal{L}(H)$ have the property $(F\text{-}P)_{\mathcal{I}}$ for some two-sided ideal $\mathcal{I}$ of $\mathcal{L}(H)$. If T in $\mathcal{I}$ is such that $AT = TB$ then for all X in $\mathcal{L}(H)$*

$$\|T - (AX - XB)\| \geq \|T\|.$$

Proof Let T in $\mathcal{I}$ be such that $AT = TB$. Let $H_1 = H = \overline{\mathrm{Ran}T} \oplus (\overline{\mathrm{Ran}T})^{\perp}$ and $H_2 = H = (\mathrm{Ker}T)^{\perp} \oplus \mathrm{Ker}T$ (Note the order). Since (A, B) has the property $(F\text{-}P)_{\mathcal{I}}$ it follows from Lemma 4.3.13 that $\overline{\mathrm{Ran}T}$ reduces A and $(\mathrm{Ker}T)^{\perp}$ reduces B. Thus, we get the decompositions of the operators

$$A = \begin{bmatrix} A_1 & 0 \\ 0 & A_2 \end{bmatrix}, \qquad B = \begin{bmatrix} B_1 & 0 \\ 0 & B_2 \end{bmatrix}.$$

For linear operators T, X from H_2 to H_1 we have

$$T = \begin{bmatrix} T_1 & 0 \\ 0 & 0 \end{bmatrix}, \qquad X = \begin{bmatrix} X_1 & X_2 \\ X_3 & X_4 \end{bmatrix},$$

say (where $T_1 : (\mathrm{Ker}T)^{\perp} \to \overline{\mathrm{Ran}T}$). Therefore,

$$\|T - (AX - XB)\| = \left\| \begin{bmatrix} T_1 - (A_1X_1 - X_1B_1) & \star \\ \star & \star \end{bmatrix} \right\|$$

$$\geq \|T_1 - (A_1X_1 - X_1B_1)\|,$$

the inequality following because the diagonal part of a block matrix always has smaller norm than that of the whole matrix. Further, $0 = AT - TB = \begin{bmatrix} A_1T_1 - T_1B_1 & 0 \\ 0 & 0 \end{bmatrix}$ forces $A_1T_1 = T_1B_1$; and since, by Lemma 4.3.13, $A_1 \left(= A|_{\overline{\mathrm{Ran}T}}\right)$ and $B_1 \left(= B|_{(\mathrm{Ker}T)^{\perp}}\right)$ are normal, Theorem 4.3.9 applies: $\|T_1 - (A_1X_1 - X_1B_1)\| \geq \|T_1\|$. Thus:

$$\|T - (AX - XB)\| \geq \|T_1 - (A_1X_1 - X_1B_1)\| \geq \|T_1\| = \|T\|.$$

$\square$

Corollary 4.3.15 *If $\{0\} \neq (\mathrm{Ker}\Delta_{A,B} \cap \mathcal{I}) \subseteq \mathrm{Ker}\Delta_{A^*,B^*}$ then $\mathrm{Ran}\Delta_{A,B} \perp (\mathrm{Ker}\Delta_{A,B} \cap \mathcal{I})$.*

An operator of the form $X^*X - XX^*$ is called a self-commutator. The subject of self-commutator approximants, that is, of minimizing the quantity

$$\|T - (X^*X - XX^*)\|$$

presents some problems. We suppose, to avoid triviality, that X is not normal. Here is a result for hyponormal X, that is, for approximation by a positive self-commutator.

Theorem 4.3.16 *Let X be hyponormal, T be isometric and $TX = XT$. Then*

$$\|T - (X^*X - XX^*)\| \geq \|T\|.$$

Proof Use the expansion (4.3.2). In (4.3.2), let "A" $= -X$, "B" $= X^*$ and take norms. Then as X is hyponormal (so that $\|X^n\| = \|X\|^n$, by Lemma 4.3.1) and T isometric we obtain (cf. proof of Theorem 4.3.3) the desired inequality. □

Observe that if X is also compact in Theorem 4.3.16 then Theorem 4.3.16 becomes the triviality $\|T\| = \|T\|$; for a compact, hyponormal operator is zero [23, Problem 206]. In Sect. 4.6 we will consider self-commutator approximation in the Von Neumann–Schatten classes C_p; there, because of the convexity properties of C_p, local arguments are available that yield more interesting results than Theorem 4.3.16.

4.4 Commutator Approximants in C_p

Here, we are concerned with minimizing the quantity

$$\|T - (AX - XA)\|_p$$

where $T - (AX - XA) \in C_p$, for $1 \leq p \leq \infty$. The structure of the Von Neumann–Schatten classes C_p is richer than that of $\mathcal{L}(H)$ due to their convexity properties. In particular, there is the uniqueness property (given in Theorem 2.4.1) and the differentiation property (given in Theorem 2.4.4). Thus, we can obtain more results in the C_p context than in $\mathcal{L}(H)$ context.

We do not first consider minimizing $\|I - (AX - XA)\|_p$ (whereas in $\mathcal{L}(H)$ we did first obtain, in Theorem 4.3.2, results on minimizing $\|I - (AX - XA)\|$); for, by Corollary 4.2.2, the operator $I - (AX - XA)$ cannot be compact.

The global result, Theorem 4.4.1, extends Anderson's theorem, Theorem 4.3.7, to C_p.

Theorem 4.4.1 *Let A be normal, $AT = TA$ and T be in C_p; let $S = \{X : AX - XA \in C_p\}$ and $F_p : S \to \mathbb{R}^+$ be given by*

$$F_p : X \mapsto \|T - (AX - XA)\|_p^p.$$

Then, for $1 \leq p < \infty$, *the map* F_p *has a global minimizer at* $X = V$ *if, and for* $1 < p < \infty$ *only if,* $AV - VA = 0$.

Proof The idea is to replace T by the compact, normal operator $|T|$. Let $T = U_1|T|$ be the polar decomposition of T so that $\mathrm{Ker}U_1 = \mathrm{Ker}|T|$ and $|T| = U_1^*T \in \mathcal{C}_p$. Since U_1 is a partial isometry so is U_1^* so that $\|U_1^*\| = 1$. As $\|U_1^*L\|_p \leq \|U_1^*\|\|L\|_p = \|L\|_p$ for arbitrary L in $\mathcal{C}_p$ (by (2.4.2)) then

$$\|T - (AX - XA)\|_p^p \geq \||T| - U_1^*(AX - XA)\|_p^p$$

$$\geq \sum_n |\langle [|T| - U_1^*(AX - XA)]\phi_n, \phi_n \rangle|_p^p, \qquad (4.4.1)$$

say, for an arbitrary orthonormal basis $\{\phi_n\}$ of the underlying space H (the last inequality following from (2.4.7)).

We construct a suitable basis $\{\phi_n\}$ that gives the desired inequality. As $AT = TA$ and as A is normal then, by Fuglede's Theorem, $AT^* = T^*A$ and hence $A|T|^2 = |T|^2A$ (for $AT^*T = T^*AT = T^*TA$). Hence, by the functional calculus, $A|T| = |T|A$ and indeed $A|T|^{p-1} = |T|^{p-1}A$. Therefore, there exists an orthonormal basis $\{\xi_k\} \cup \{\psi_m\}$ of H such that $\{\psi_m\}$ is an orthonormal basis of $\mathrm{Ker}|T|$ and $\{\xi_k\}$ consists of common eigenvectors of (the commuting normal operators) A and $|T|$. Hence, $\sum_k < |T|\xi_k, \xi_k >^p = \|T\|_p^p$. In (4.4.1) take $\{\phi_n\} = \{\xi_k\} \cup \{\psi_m\}$. As $AT^* = T^*A$ and $A|T| = |T|A$ we get $|T|U_1^*A = T^*A = AT^* = A|T|U_1^* = |T|AU_1^*$. Let $|T|\xi_k = t_k\xi_k$, say, where $t_k > 0$. For the R.H ξ_k in $< U_1^*AX\xi_k, \xi_k >$ substitute $\xi_k = \frac{1}{t_k}|T|\xi_k$. Then

$$\langle U_1^*AX\xi_k, \xi_k \rangle = \frac{1}{t_k}\langle |T|U_1^*AX\xi_k, \xi_k \rangle = \frac{1}{t_k}\langle |T|AU_1^*X\xi_k, \xi_k \rangle = \langle AU_1^*X\xi_k, \xi_k \rangle.$$

Hence, as ξ_k is also an eigenvector of the normal operator A (with $A\xi_k = \alpha_k\xi_k$, say) then

$$\langle U_1^*(AX - XA)\xi_k, \xi_k \rangle = \langle AU_1^*X\xi_k, \xi_k \rangle - \langle U_1^*XA\xi_k, \xi_k \rangle$$

$$= \alpha_k \langle U_1^*X\xi_k, \xi_k \rangle - \alpha_k \langle U_1^*X\xi_k, \xi_k \rangle = 0.$$

Returning now to (4.4.1):

$$\|T - (AX - XA)\|_p^p \geq \sum_k \langle |T|\xi_k, \xi_k \rangle^p + \sum_m |\langle U_1^*(AX - XA)\psi_m, \psi_m \rangle|^p$$

$$= \sum_k \langle |T|\xi_k, \xi_k \rangle^p = \|T\|_p^p$$

as desired (the second equality following because $\psi_m \in \mathrm{Ker}|T| = \mathrm{Ker}U_1$).

For $1 < p < \infty$, the uniqueness assertion follows, by Theorem 2.4.1, from the convexity of the set $\mathcal{S} = \{X : AX - XA \in \mathcal{C}_p\}$. $\qquad\square$

In other words, under the hypotheses of Theorem 4.4.1,

$$\|T - (AX - XA)\|_p \geq \|T\|_p, \quad 1 \leq p < \infty \qquad (4.4.2)$$

with equality if, and for $1 < p < \infty$ only if, $AX - XA = 0$.

Geometrically, in terms of the operator Δ_A, the inequality (4.4.2) says that $\mathrm{Ran}\,\Delta_{A|\mathcal{C}_p}$ is orthogonal to $\mathrm{Ker}\,\Delta_{A|\mathcal{C}_p}$, where $A|\mathcal{C}_p$ denotes the restriction of A to $\mathcal{C}_p$ (cf. Corollary 4.3.4, 4.3.8). With $\overline{\mathrm{Ran}\,\Delta_{A|\mathcal{C}_p}}$ denoting the norm closure of the range of $\Delta_{A|\mathcal{C}_p}$ in $\mathcal{C}_p$, this orthogonality may be expressed by

$$\overline{\mathrm{Ran}\,\Delta_{A|\mathcal{C}_p}} \cap \mathrm{Ker}\,\Delta_{A|\mathcal{C}_p} = \{0\} \qquad (4.4.3)$$

The reasoning of Theorem 4.4.1 gives a simple proof of Anderson's Theorem (Theorem 4.3.7) in the special case when T is compact.

Corollary 4.4.2 Let A be normal, $AT = TA$ and T be compact. Then

$$\|T - (AX - XA)\| \geq \|T\|.$$

Proof Using the fact that $\|L\| = \sup_{\|\phi\|=1} |\langle L\phi, \phi\rangle|$, where $L \in \mathcal{L}(H)$, with the basis $\{\phi_n\} = \{\xi_k\} \cup \{\psi_m\}$ and the partial isometry U_1 as defined in the proof of Theorem 4.4.1 we have

$$\|T - (AX - XA)\| \geq \||T| - U_1^*(AX - XA)\|$$
$$\geq \sup_n |\langle [|T| - U_1^*(AX - XA)]\phi_n, \phi_n\rangle|$$
$$= \sup_{k,m} \left[\langle |T|\xi_k, \xi_k\rangle + |\langle U_1^*(AX - XA)\psi_m, \psi_m\rangle|\right]$$
$$= \sup \langle |T|\xi_k, \xi_k\rangle$$
$$= \||T|\| = \|T\|,$$

the penultimate equality following because T, hence $|T|$, is compact. $\qquad\square$

We make some comments about the global results above.

1. In Theorem 4.4.1, (4.4.2) and Corollary 4.4.2 if $T = AX_1 - X_1A$ for some operator X_1 then the minimum of $\|T - (AX - XA)\|_p$, for $1 \leq p \leq \infty$, is 0. This does not conflict with the global results above because in this case $T = 0$; for since the normal operator A commutes with $T(= AX_1 - X_1A)$ then, by Theorem 4.2.4, $AX_1 - X_1A = 0$.

2. The following counter-example shows that Theorem 4.4.1 does not hold if $p < 1$. Take $p = \frac{1}{2}$ and

$$
A = \begin{bmatrix} a & 0 \\ 0 & 0 \end{bmatrix}, \quad T = \begin{bmatrix} 1 & 0 \\ 0 & 1 \end{bmatrix} \quad \text{and} \quad X = \begin{bmatrix} 0 & -x \\ x & 0 \end{bmatrix},
$$

where a and x are reals such that $0 < |ax| < 1$. Then $\|T - (AX - XA)\|_{\frac{1}{2}} < \|T\|_{\frac{1}{2}}$.

The local results obtained in this section, and in the next, hinge on the following lemma. The proof of the Lemma 4.4.3 uses ideas about the geometry of partial isometries.

Lemma 4.4.3 *Let A and B be normal and $W = U|W|$ be the polar decomposition of W. If, for $1 < p < \infty$,*

$$
A|W|^{p-1}U^* = |W|^{p-1}U^*B \tag{4.4.4}
$$

then

$$
A|W|U^* = |W|U^*B \tag{4.4.5}
$$

Proof Write $Z = |W|^{p-1}$. Then (4.4.4) says that

$$
AZU^* = ZU^*B \tag{4.4.6}
$$

and (4.4.5) says that $AZ^{\frac{1}{p-1}}U^* = Z^{\frac{1}{p-1}}U^*B$. This will follow by the functional calculus from

$$
AZ^nU^* = Z^nU^*B; \tag{4.4.7}
$$

where $n \in \mathbb{N}$. In the polar decomposition $W = U|Z|$ we have $\text{Ker}U = \text{Ker}|W| = \text{Ker}Z$ so that U^*U, the projection onto $(\text{Ker}U)^\perp = \text{Ran}Z$ satisfies $ZU^*UZ = Z^2$. From (4.4.6) it follows by Putnam's Corollary, Corollary 4.3.11, that $A^*(ZU^*) = (ZU^*)B^*$ (since A and B are normal), that is, $UZA = BUZ$. Therefore, from (4.4.6) we get $AZ^2 = AZU^*UZ = ZU^*BUZ = ZU^*UZA = Z^2A$. Hence, $AZ = ZA$. The equality (4.4.7) now follows by induction. $\quad\square$

The upshot of the local result, Theorem 4.4.4 below, is that V is a critical point of the map $X \mapsto \|T - (AX - XA)\|_p^p$, where $1 < p < \infty$, if and only if it is a global minimizer of it.

Theorem 4.4.4 *Let A be normal, $AT = TA$ and T be in C_p; let $S = \{X : AX - XA \in C_p\}$ and let $F_p : S \to \mathbb{R}^+$ be given by*

$$
F_p : X \mapsto \|T - (AX - XA)\|_p^p.
$$

Then:

(a) *for $1 < p < \infty$, the map F_p has a critical point at $X = V$ if and only if $AV - VA = 0$;*

(b) *for $0 < p \le 1$, the map F_p has a critical point at $X = V$ if $AV - VA = 0$ provided $\dim H < \infty$ and $T - (AV - VA)$ is invertible.*

Proof

(a) Let V be in $\mathcal{S}$ so that $T - (AV - VA) \in \mathcal{C}_p$. Let S be an arbitrary increment of V subject to $T - (A(V + S) - (V + S)A) \in \mathcal{C}_p$, that is, $SA - AS \in \mathcal{C}_p$ (so that $S \in \mathcal{S}$). Let $\Psi : X \mapsto T - (AX - XA)$ and $\Phi : X \mapsto \|X\|_p^p$ so that $F_p = \Phi \circ \Psi$. Let $D_V F_p$ denote the Fréchet derivative of F_p at V. Then

$$(D_V F_p)(S) = D_{(T-(AV-VA))}\Phi(SA - AS). \tag{4.4.8}$$

Let $T - (AV - VA) = U|T - (AV - VA)|$ be the polar decomposition of $T - (AV - VA)$. Then from (4.4.8) and by Theorem 2.4.4(a)

$$(D_V F_p)(S) = p\mathcal{R}\tau[|T - (AV - VA)|^{p-1}U^*(SA - AS)]$$
$$= p\mathcal{R}\tau[Y(SA - AS)]$$
$$= p\mathcal{R}\tau[(AY - YA)S] \tag{4.4.9}$$

where $Y = |T-(AV-VA)|^{p-1}U^*$ and where $\tau[YSA] = \tau[AYS]$ by the invariance of trace (Sect. 2.4, [40, Theorem 2.2.4 (iv)]).

Let V be a critical point of F_p so that $(D_V F_p)(S) = 0$ for all S in $\mathcal{L}(H)$. The proof now follows similar lines to that of Theorem 3.2.4. Take $S = f \otimes g$ where f and g are arbitrary vectors in H and where $f \otimes g : x \to \langle x, f \rangle g$. As $\tau[M(f \otimes g)] = \langle Mg, f \rangle$ for arbitrary M in $\mathcal{L}(H)$, by (2.4.5), then substituting in (4.4.9) we get

$$0 = \mathcal{R}\tau[(AY - YA)S] = \mathcal{R}\langle(AY - YA)g, f\rangle.$$

As f and g are arbitrary this forces $AY - YA = 0$, that is,

$$A|T - (AV - VA)|^{p-1}U^* = |T(AV - VA)|^{p-1}U^*A.$$

Lemma 4.4.3 applies with $A(= \text{``}B\text{''})$ normal we have

$$A|T - (AV - VA)|U^* = |T - (AV - VA)|U^*A.$$

This says that $A(T-(AV-VA))^* = (T-(AV-VA))^*A$. Since $AT = TA$ then, by Fuglede's Theorem, $AT^* = T^*A$ and so $A(AV - VA)^* = (AV - VA)^*A$. Taking adjoints and using Fuglede's Theorem again we get $A(AV - VA) = (AV - VA)A$. Therefore, by Theorem 4.2.4, $AV - VA = 0$.

Conclusion so far: V is a critical point of $F_p \implies AV - VA = 0$.

To prove the converse implication let V satisfy $AV - VA = 0$. Then $T - (AV - VA) = T$ and so the partial isometries U and U_1 say, in the polar decomposition of $T - (AV - VA)$ and T coincide. Thus, $Y = |T|^{p-1}U_1^* \in \mathcal{C}_1$. As in the proof of Theorem 4.4.1 since the normal operator A commutes with T then, by Fuglede, $AT^* = T^*A$ and so $A|T| = |T|A$ and indeed $A|T|^{p-1} = |T|^{p-1}A$. Hence also $|T|U^*A = |T|AU^*$. Thus, $\text{Ran}(AU^* - U^*A) \subseteq \text{Ker}|T| = \text{Ker}|T|^{p-1}$. Therefore,

$$AY - YA = A|T|^{p-1}U^* - |T|^{p-1}U^*A = |T|^{p-1}(AU^* - U^*A) = 0.$$

So, as $YS \in \mathcal{C}_1$ then by (4.4.9) we have $(D_V F_p)(S) = 0$ for all S in $\mathcal{L}(H)$, that is, V is a critical point of F_p.

(b) For $0 < p \leq 1$, the finite dimensionality and invertibility conditions, ensure, by Theorem 2.4.4(b), that F_p is differentiable at V. If $AV - VA = 0$ then T, and hence $|T|$ is invertible and so $|T|^{p-1}$ exists for $0 < p \leq 1$. The proof of the implication, $AV - VA = 0 \implies V$ is a critical point of F_p, is now the same as in part (a).   □

We make two comments about the local result.

1. Of course, the converse in Theorem 4.4.4(a) can be proved on the basis of the global result : if $AV - VA = 0$ then, by Theorem 4.4.1, V is a global minimizer of F_p, for $1 \leq p < \infty$, and hence a critical point of it, for $1 < p < \infty$.
2. The proof in Theorem 4.4.4(a) of the implication, V is a critical point of $F_p \implies AV - VA = 0$, does not work in the $0 < p \leq 1$ case, because the functional calculus argument in Lemma 4.4.3 involving the function $f : t \to t^{\frac{1}{p-1}}$, where $0 \leq p < \infty$, is valid only for $1 < p < \infty$.

4.5 Generalized Commutator Approximants in $\mathcal{C}_p$

We extend the analysis to approximation, in $\mathcal{C}_p$, by generalized commutators, that is, operators of the form $AX - XB$. First, we have the following version of Theorem 4.3.9.

Theorem 4.5.1 *Let A and B be normal, $AT = TB$ and T be in $\mathcal{C}_p$; let $S = \{X : AX - XB \in \mathcal{C}_p\}$ and $F_p : S \to \mathbb{R}^+$ be given by*

$$F_p : X \mapsto \|T - (AX - XB)\|_p^p.$$

Then, for $1 \leq p < \infty$, the map F_p has a global minimizer of $X = V$ if, and for $1 < p < \infty$ only if, $AV - VB = 0$.

Proof Notation apart (here, $\| \cdot \|_p$ replaces $\| \cdot \|$), the proof of the minimization result is the same as that of Theorem 4.3.9; just as Theorem 4.3.9 is obtained from Theorem 4.3.7 by operator matrices so, too, the result here is obtained from Theorem 4.4.1. The proof of the equality assertion is the same as in Theorem 4.4.1.

□

Thus, under the hypotheses of the Theorem 4.5.1,

$$\|T - (AX - XB)\|_p \geq \|T\|_p, \quad 1 \leq p < \infty \tag{4.5.1}$$

with equality if, and for $1 < p < \infty$ only if, $AX - XB = 0$.

We state the $\mathcal{C}_p$ Fuglede–Putnam global inequality, Theorem 4.5.2. We omit the proof of Theorem 4.5.2 as it is the same as its $\mathcal{L}(H)$ counterpart, Theorem 4.3.14.

Theorem 4.5.2 *Let the pair (A, B) in $\mathcal{L}(H)$ have the property $(F\text{-}P)_{\mathcal{C}_p}$, where $1 \leq p < \infty$, let T in $\mathcal{C}_p$ be such that $AT = TB$ and let $\mathcal{S} = \{X : AX - XB \in \mathcal{C}_p\}$. Then for all X in $\mathcal{S}$*

$$\|T - (AX - XB)\|_p \geq \|T\|_p$$

with equality if, and for $1 < p < \infty$ only if, $AX - XB = 0$.

Thus, under the hypotheses of Theorem 4.5.2, $\mathrm{Ran}\Delta_{A,B|\mathcal{C}_p}$ is orthogonal to $\mathrm{Ker}\Delta_{A,B|\mathcal{C}_p}$ (Recall $\Delta_{A,B}(X) = AX - XB$ and $\Delta_{A,B|\mathcal{C}_p}$ denotes the restriction of $\Delta_{A,B}$ to $\mathcal{C}_p$); that is,

$$\overline{\mathrm{Ran}\Delta_{A,B|\mathcal{C}_p}} \cap \mathrm{Ker}\Delta_{A,B|\mathcal{C}_p} = \{0\} \tag{4.5.2}$$

where $\overline{\mathrm{Ran}\Delta_{A,B|\mathcal{C}_p}}$ denotes the norm closure of the range of $\Delta_{A,B|\mathcal{C}_p}$.

Theorem 4.5.3 *Let A and B be normal, $AT = TB$ and T be in $\mathcal{C}_p$; let $\mathcal{S} = \{X : AX - XB \in \mathcal{C}_p\}$ and let $F_p : \mathcal{S} \to \mathbb{R}^+$ be given by*

$$F_p : X \mapsto \|T - (AX - XB)\|_p^p.$$

Then:

(a) *for $1 < p < \infty$, the map F_p has a critical point at $X = V$ if and only if $AV - VB = 0$;*

(b) *for $0 \leq p < 1$, the map F_p has a critical point at $X = V$ if $AV - VB = 0$ provided $\dim H < \infty$ and $T - (AV - VB)$ is invertible.*

Proof

(a) Let V be in $\mathcal{S}$ and S be an arbitrary increment of V such that $T - (A(V+S)) - (V+S)B) \in \mathcal{C}_p$, that is, $SB - AS \in \mathcal{C}_p$. Let $\Phi(X) = \|X\|_p^p$ where $X \in \mathcal{C}_p$. Let $D_V F_p$ denote the Fréchet derivative of F_p at V. Then, cf. the proof of Theorem 4.4.4,

$$(D_V F_p)(S) = D_{(T-(AV-VB))}\Phi(SB - AS).$$

Let $T - (AV - VB) = U|T - (AV - VB)|$ be the polar decomposition of $T -$
$(AV - VB)$ and $Y = |T - (AV - VB)|^{p-1}U^*$. Then, by Theorem 2.4.4

$$(D_V F_p)(S) = p\mathcal{R}\tau[Y(SB - AS)]$$
$$= p\mathcal{R}\tau[(BY - YA)S]$$

by the invariance of trace (Sect. 2.4).

Let V be a critical point of F_p. Then, cf. the proof of Theorem 4.4.4, it follows
that $BY - AY = 0$, that is,

$$B|T - (AV - VB)|^{p-1}U^* = |T - (AV - VB)|^{p-1}U^*A.$$

Since A and B are normal then, by Lemma 4.4.3,

$$B|T - (AV - VB)|U^* = |T - (AV - VB)|U^*A. \qquad (4.5.3)$$

As $AT = TB$ then $A^*T = TB^*$ by Putnam's corollary, Corollary 4.3.11.
Therefore, from (4.5.3) $B(AV - VB)^* = (AV - VB)^*A$ so that on taking
adjoints and applying Putnam's corollary to the normal operators A^*, B^* we
get $A(AV - VB) - (AV - VB)B = 0$. Thus, by (4.5.2),

$$AV - VB \in \operatorname{Ran}\Delta_{A,B|\mathcal{C}_p} \cap \operatorname{Ker}\Delta_{A,B|\mathcal{C}_p} = \{0\}.$$

Hence, $AV - VB = 0$.

Conversely, if $AV - VB = 0$ then, by Theorem 4.5.1, V is a global minimizer
of F_p, for $1 \le p < \infty$, and hence, for $1 < p < \infty$, a critical point of F_p.
(b) The proof is omitted since it is the same as that of Theorem 4.4.4(b).   □

4.6   Self-Commutator Approximants in $\mathcal{C}_p$

In minimizing the quantity

$$\|T - (X^*X - XX^*)\|_p, \quad 1 \le p < \infty$$

the problem is to frame suitable hypotheses: clearly, X cannot be normal nor,
as commented in Sect. 4.3, be compact and hyponormal. Now, the results on

commutator approximation, Theorems 4.3.7 and 4.4.1, on minimizing $\|T - (AX - XA)\|_p$ for $1 \leq p \leq \infty$ say that the zero commutator $(0 = AX - XA)$ is a commutator approximant of T (similarly for generalized commutator approximation: see Theorems 4.3.9, 4.3.14, 4.5.1 and 4.5.2).

Since a global minimizer is a critical point we study the local behaviour of the map

$$F_p : X \mapsto \|T - (X^*X - XX^*)\|_p^p, \quad 1 < p < \infty.$$

We show in Theorem 4.6.1 that if a critical point (hence in particular, a global minimizer) V of F_p satisfies $V^*V - VV^* = 0$ for self-adjoint T then $TV = VT$. Theorem 4.6.2 says that for self-adjoint T such that $TX = XT$ the point V is a critical point if and only if $V^*V - VV^* = 0$.

It follows from Theorem 4.6.2 that every global minimizer X of F_p satisfies $X^*X - XX^* = 0$ for self-adjoint T commuting with X. The global result, Theorem 4.6.3, guarantees the **existence** of global minima. Thus, it says that under the same hypotheses $(TX = XT, T^* = T \in \mathcal{C}_p)$ for all X such that $X^*X - XX^* \in \mathcal{C}_p$, where $1 < p < \infty$,

$$\|T - (XX^* - XX^*)\|_p \geq \|T\|_p \tag{4.6.1}$$

with equality in (4.6.1) if and only if $X^*X - XX^* = 0$.

Thus, the approach here for $1 < p < \infty$ is quite different to that of Sects. 4.4, 4.5: there, the local and global results are independent of each other; here, the statement—and proof—of the global result, Theorem 4.6.3, depends on the local results Theorems 4.6.1, 4.6.2.

There is a similar inequality for the trace norm $(p = 1)$, Theorem 4.6.4. Theorem 4.6.4 cannot, of course, be deduced from local considerations. Instead, the proof of Theorem is modelled on that of Theorem 4.4.1.

Examples 4.6.6–4.6.8 illustrate, and reinforce, the results. Examples 4.6.6 and 4.6.7 show that, if the (seemingly restrictive) condition $TX = XT$ is dropped, the conclusions of Theorems 4.6.3 and 4.6.4 do not hold. Example 4.6.8 shows that for $0 < p < 1$ the inequalities may be reversed.

Theorem 4.6.1 *Let T be self-adjoint and in $\mathcal{C}_p$ where $1 < p < \infty$. Let $\mathcal{S} = \{X : X^*X - XX^* \in \mathcal{C}_p\}$ and let $F_p : \mathcal{S} \to \mathbb{R}^+$ be given by*

$$F_p : X \mapsto \|T - (X^*X - XX^*)\|_p^p.$$

Then if V is a critical point of F_p such that $V^*V - VV^* = 0$ it follows that $TV = VT$.

Proof As we shall use this proof in that of Theorem 4.6.2 we shall adopt the hypothesis that $V^*V - VV^* = 0$ only at the last step.

Step 1 Let V be in $\mathcal{S}$ so that $T - (V^*V - VV^*) \in \mathcal{C}_p$. (Observe that $\mathcal{S}$ properly contains $\mathcal{C}_p$ for if $X \in \mathcal{C}_p$ then $X \in \mathcal{S}$ and, e.g., $I \in \mathcal{S}$ but $I \notin \mathcal{C}_p$.) Let $\mathcal{F}$ consist of all operators S for which $T - [(V + S)^*(V + S) - (V + S)(V + S)^*] \in \mathcal{C}_p$ (thus, $\mathcal{F}$ also properly contains $\mathcal{C}_p$). Let $\Phi : X \mapsto \|X\|_p^p$. By considering $F_p(V + S) - F_p(V)$, for arbitrary S in $\mathcal{F}$, it follows that

$$(D_V F_p)(S) = D_{T-(V^*V-VV^*)}\Phi(VS^* + SV^* - V^*S - S^*V).$$

Let $T - (V^*V - VV^*) = U|T - (V^*V - VV^*)|$ be the polar decomposition of $T - (V^*V - VV^*)$. Then, by Theorem 2.4.4, on writing $Y = U|T - (V^*V - VV^*)|^{p-1}$,

$$(D_V F_p)(S) = p\mathcal{R}\tau[Y^*(VS^* + SV^* - V^*S - S^*V)] \qquad (4.6.2)$$

for all operators S in $\mathcal{F}$. Note that $Y^*(= |T - (V^*V - VV^*)|^{p-1}U^*) \in \mathcal{C}_1$. Therefore, as $\mathcal{R}\tau[A] = \mathcal{R}\tau[A^*]$ for all $A \in \mathcal{C}_1$, we have $\mathcal{R}\tau[Y^*VS^* - Y^*S^*V] = \mathcal{R}\tau[SV^*Y - V^*SY]$. Hence, by the invariance of trace,

$$(D_V F_p)(S) = p\mathcal{R}\tau[(V^*Y - YV^* + V^*Y^* - Y^*V^*)S]. \qquad (4.6.3)$$

Step 2 Let V be a critical point of F_p so that $(D_V F_p)(S) = 0$ for all S in $\mathcal{F}$. Then, reasoning as in the proof of Theorems 4.4.4(a), we find from (4.6.3) that $V^*Y - YV^* + V^*Y^* - Y^*V^* = 0$, that is,

$$(\mathcal{R}Y)V = V(\mathcal{R}Y). \qquad (4.6.4)$$

Step 3 Let T be self-adjoint. Then $T - (V^*V - VV^*)(= U|T - (V^*V - VV^*)|)$ is self-adjoint. Hence, U is self-adjoint and commutes with $|T - (V^*V - VV^*)|$ and hence $Y(= U|T - (V^*V - VV^*)|^{p-1})$ is self-adjoint. Hence, from (4.6.4), $YV = VY$, that is

$$U|T - (V^*V - VV^*)|^{p-1}V = VU|T - (V^*V - VV^*)|^{p-1} \qquad (4.6.5)$$

Step 4 Assertion: V satisfies

$$U|T - (V^*V - VV^*)|V = VU|T - (V^*V - VV^*)| \qquad (4.6.6)$$

Proof of Assertion (Sketch): Let $Z = |T - (V^*V - VV^*)|^{p-1}$ so that, by (4.6.5), $UZV = VUZ$ and, as $Y = UZ$ is self-adjoint and $ZU = UZ$, $ZU^*V = VZU^*$ (it is simplest to write, where necessary, U^*); $Z^2 = ZU^*UZ$ (as in Lemma 4.4.3), therefore $Z^2V = (ZU^*V)UZ = VZ^2$ whence (reasoning as in Lemma 4.4.3) the equality (4.6.6) follows.

The equality (4.6.6) says that

$$TV - (V^*V - VV^*)V = VT - V(V^*V - VV^*). \qquad (4.6.7)$$

Step 5 If, finally, $V^*V - VV^* = 0$ then (4.6.7) forces $TV = VT$.   □

Note

(1) As in Lemma 4.4.3, the proof of the assertion holds only for $1 < p < \infty$.
(2) Observe that for non-self-adjoint T, in the case $p = 2$, it follows from equality
     (4.6.4) of the proof of Theorem 4.6.1 that if V is a critical point of F_p such that
     $V^*V - VV^* = 0$ then $(\mathcal{R}T)V = V(\mathcal{R}T)$. But this later equality does not force
     $TV = VT$, even if T is normal: witness $T = \begin{bmatrix} 1 & i \\ i & 1 \end{bmatrix}$.

Theorem 4.6.2 *Let T be self-adjoint, let $TX = XT$ and T be in C_p. Let $S = \{X :$
$X^*X - XX^* \in C_p\}$ and let $F_p : S \to \mathbb{R}^+$ be given by*

$$F_p : X \mapsto \|T - (X^*X - XX^*)\|_p^p.$$

Then:

*(a) for $1 < p < \infty$, the map F_p has a critical point at $X = V$ if and only if
     $V^*V - VV^* = 0$;*
*(b) for $0 < p \le 1$, the map F_p has a critical point at $X = V$ if $V^*V - VV^* = 0$
     provided $\dim H < \infty$ and $T - (V^*V - VV^*)$ is invertible;*
*(c) for $p = 2$, the same result as in (a) holds if the condition on T of self-adjointness
     is replaced by normality.*

Proof

(a) Let V be a critical point of F_p. The equality (4.6.7) of the proof of Theorem 4.6.1
     holds so that, as $TV = VT$, then $(V^*V - VV^*)V = V(V^*V - VV^*)$, whence by
     Kleinecke–Shirokov (Theorem 4.2.3) $V^*V - VV^*$ is quasinilpotent and hence,
     being self-adjoint, zero.
       Conversely, let V satisfy $V^*V - VV^* = 0$. Then (cf. proof of the converse
     assertion in Theorem 4.4.4(a)), the partial isometries U and, say, U_1 occurring
     in the polar decompositions of $T - (V^*V - VV^*)$ and of T coincide. Thus,
     $Y = U|T|^{p-1} \in C_1$ so that $Y^* = |T|^{p-1}U^* \in C_1$.
       We first prove that $Y^*V - VY^* = 0$. Since V and V^* commute with T they
     commute with $|T|$ (and hence with $|T|^{p-1}$). So, $|T|U^*V = |T|VU^*$. Hence,
     $\mathrm{Ran}(U^*V - VU^*) \subseteq \mathrm{Ker}|T| = \mathrm{Ker}|T|^{p-1}$. Therefore, since $|T|^{p-1}V = V|T|^{p-1}$
     it follows that $Y^*V - VY^* = 0$.
       Similarly, from the equality $|T|U^*V^* = |T|V^*U^*$ it follows that $Y^*V^* -$
     $V^*Y^* = 0$. Hence, $V^*Y - YV^* + V^*Y^* - Y^*V^* = 0$. Substitute into equality
     (4.6.3) of the proof of Theorem 4.6.1 (the expression for $(D_V F_p)(S)$). As $YS \in$
     C_1 and $Y^*S \in C_1$ it follows from (4.6.3) that $(D_V F_p)(S) = 0$ for all S in $\mathcal{L}(H)$.
(b) Follows immediately from (a) as in Theorem 4.4.4(b).

(c) Let T be normal. If V is a critical point of F_2, then equality (4.6.4) of the proof of Theorem 4.6.1 says that

$$[\mathcal{R}T - (V^*V - VV^*)]V = V[\mathcal{R}T - (V^*V - VV^*)].$$

Since V commutes with T then, by Fuglede, V commutes with $\mathcal{R}T$. The result now follows, via Kleinecke–Shirokov, as in (a).

The proof of the converse implication (V satisfies $V^*V - VV^* = 0 \implies V$ is a critical point of F_2) depends only on V and V^* commuting with T and is therefore the same as in (a).   □

Indeed, the proof in (a) of the implication, V satisfies $V^*V - VV^* = 0 \implies V$ is a critical point of F_p, for $1 < p < \infty$ holds (via Fuglede) for normal T commuting with V.

Theorem 4.6.3 *Let T be self-adjoint, let $TX = XT$ and let T be in $\mathcal{C}_p$. Let $\mathcal{S} = \{X : X^*X - XX^* \in \mathcal{C}_p\}$. Then, if $X \in \mathcal{S}$,*

(a) for $1 < p < \infty$,

$$\|T - (X^*X - XX^*)\|_p \geq \|T\|_p \tag{4.6.8}$$

with equality holding in (4.6.8) if and only if $X^*X - XX^* = 0$;

(b) for $p = 2$, the same result as in (a) holds if T is assumed normal rather than self-adjoint.

Proof

(a) First, suppose the operators X in $\mathcal{S}$ are contractions, i.e such that $\|X\| \leq 1$. Suppose also that the underlying space H is finite-dimensional (a similar argument will be used in Theorem 6.3.10). The set of contractions is bounded and closed (for the condition $X^*X - I \leq 0$ characterises the contractions and the map $X \mapsto X^*X$ is continuous [23, Problem 129]). Hence, $\mathcal{S}$ is compact since H is finite-dimensional. Therefore, the continuous map $F_p : X \mapsto \|T - (X^*X - XX^*)\|_p^p$ is bounded, attains its bounds and thus has a global minimizer and hence a critical point at $X = V$, say. Since, by Theorem 4.6.2(a), $V^*V - VV^* = 0$ then

$$\|T - (X^*X - XX^*)\|_p \geq \|T\|_p.$$

Conversely, if equality holds in (4.6.8) for some point X, then that point X is a global minimizer, hence a critical point, of F_p, whence, by Theorem 4.6.2(a), $X^*X - XX^* = 0$.

The extension to infinite-dimensional H is similar to [1, Theorem 3.5]. As the operator T is compact and normal there exists a basis $\{\phi_i\}$ of H consisting of eigenvectors of T which may be ordered such that $|\lambda_1| \geq |\lambda_2| \geq \ldots$, where $T\phi_i = \lambda_i\phi_i$ (and where the eigenvalues are repeated according to multiplicity).

Let

$$H_k = \text{span}\{\phi_i : T\phi_i = \lambda_i\phi_i, i = 1, \ldots, k\}.$$

Then H_k is invariant under X and X^*; for if ϕ_i is an eigenvector of T then so are $X\phi_i$ and $X^*\phi_i$ with the same eigenvalues (since T commutes with X and X^*). Therefore, if E_k denotes the orthogonal projection onto H_k then $E_kX = XE_k$. Hence, E_kTE_k commutes with E_kXE_k (and with $E_kX^*E_k$) and hence by the finite-dimensional inequality (4.6.8) applied to the contraction E_kXE_k we have

$$\|(E_kTE_k) - [(E_kXE_k)^*(E_kXE_k) - (E_kXE_k)(E_kXE_k)^*]\|_p \geq \|E_kTE_k\|_p,$$

that is, $\|E_k[T - (X^*X - XX^*)]E_k\|_p \geq \|E_kTE_k\|_p$. Now, let $k \to \infty$ then $E_k \to I$ and from [15, Lemma 2] (cf. [1, Theorem 3.5]) it follows that the inequality (4.6.8) holds for infinite-dimensional H.

The condition that the operator X in $\mathcal{S}$ is a contraction may now be lifted. Let X be arbitrary in $\mathcal{S}$; then by applying the inequality (4.6.8) to the contraction $\frac{X}{\|X\|}$ the result immediately follows.

(b) follows immediately to (a) from the corresponding local result, Theorem 4.6.2(c). □

Theorem 4.6.4 *Let T be normal, let $TX = XT$ and T be in $\mathcal{C}_1$. Then, if $X^*X - XX^* \in \mathcal{C}_1$*

$$\|T - (X^*X - XX^*)\|_1 \geq \|T\|_1.$$

Proof Let $U|T|$ be the polar decomposition of T. As U is a partial isometry so is U^* and so $\|U^*\| = 1$. Since $\|U^*L\|_1 \leq \|U^*\|\|L\|_1 = \|L\|_1$ for arbitrary L in $\mathcal{C}_1$ (by Eq. (2.4.2)) then

$$\|T - (X^*X - XX^*)\|_1 \geq \||T| - U^*(X^*X - XX^*)\|_1$$
$$\geq |\tau[|T| - U^*(X^*X - XX^*)]|$$

by Ringrose [40, Lemma 2.3.3] where

$$\tau[|T| - U^*(X^*X - XX^*)] = \sum_n \langle [|T| - U^*(X^*X - XX^*)]\phi_n, \phi_n \rangle \qquad (4.6.9)$$

for an arbitrary orthonormal basis $\{\phi_n\}$ of H.

Take $\{\phi_n\}$ as the orthonormal basis of H consisting of eigenvectors of the compact normal operator $|T|$. Let $\{\psi_m\}$ be an orthonormal basis of $\text{Ker}|T|$ and let $\{\xi_k\}$ be an orthonormal basis of $(\text{Ker}|T|)^\perp$ consisting of eigenvectors of $|T|$. Thus, $\{\phi_n\} = \{\psi_m\} \cup \{\xi_k\}$. Then $\sum_m \langle [|T| - U^*(X^*X - XX^*)]\psi_m, \psi_m \rangle = 0$ because $\psi_m \in \text{Ker}|T| = \text{Ker}U$; and $\sum_k \langle |T|\xi_k, \xi_k \rangle = \|T\|_1$. Further, since $TX = XT$ and

$TX^* = X^*T$ (by Fuglede) then $|T|X^* = X^*|T|$, and so $|T|U^*X^*X = T^*X^*X = X^*T^*X = X^*|T|U^*X = |T|X^*U^*X$. Then reasoning as in Theorem 4.4.1, we find that $\langle U^*X^*X\xi_k, \xi_k \rangle = \langle X^*U^*X\xi_k, \xi_k \rangle$. By the invariance of trace (over $(\mathrm{Ker}|T|)^\perp$), $\sum_k \langle U^*XX^*\xi_k, \xi_k \rangle = \sum \langle X^*U^*X\xi_k, \xi_k \rangle$. Thus, from (4.6.9)

$$\|T - (X^*X - XX^*)\|_1 \geq \sum_k \langle |T|\xi_k, \xi_k \rangle = \|T\|_1.$$

□

In the case where T is positive the proof of the trace norm result is simple and does not require the commutativity condition.

Theorem 4.6.5 *Let T be positive and be in C_1. Then, if $X^*X - XX^* \in C_1$,*

$$\|T - (X^*X - XX^*)\|_1 \geq \|T\|_1.$$

Proof Let T be positive so that $T = |T|$. Then from the well-known inequality $\|Y\| \geq |\tau(Y)|$ for each Y in C_1 [40, Lemma 2.3.3] and from the linearity and invariance of trace we have

$$\|T - (X^*X - XX^*)\|_1 \geq |\tau[T - (X^*X - XX^*)]|$$
$$= |\tau[T]| = \tau[|T|] = \|T\|_1.$$

□

If T and X do not commute then either the inequality $\|T-(X^*X-XX^*)\|_p \geq \|T\|_p$, for $1 \leq p < \infty$, is reversed (Example 4.6.6) or $\|T-(X^*X-XX^*)\|_p = \|T\|_p$ without $X^*X - XX^* = 0$ (Example 4.6.7).

Example 4.6.6 Take $T = \begin{bmatrix} 3 & 0 \\ 0 & -3 \end{bmatrix}$ and $X = \begin{bmatrix} 1 & 1 \\ 2 & 2 \end{bmatrix}$ so that $T = T^*$ and $TX \neq XT$.

Recall the definition of the $\|\cdot\|_p$ norm (Eq. (2.4.1)): $\|A\|_p^p = \sum_p s_i^p(A)$ where $s_i(A)$ denotes the ith eigenvalue of $|A|$, where $A \in C_p$. Here,

$$\|T - (X^*X - XX^*)\|_p^p = 1^p + 1^p < 3^p + 3^p = \|T\|_p^p, \quad 1 \leq p < \infty.$$

Example 4.6.7 Take $X = f \otimes g$ and $T = f \otimes f$ where $f \neq g$ and $\|f\| = \|g\|$. Then it can be checked (from Eq. (2.3.1)) that $T = T^* \geq 0$, that $TX \neq XT$ and that $X^*X - XX^* = f \otimes f - g \otimes g \neq 0$. Further, since $\|f \otimes g\|_p = \|f\| \|g\|$ for $1 \leq p < \infty$ [40, p. 90] we have

$$\|T - (X^*X - XX^*)\|_p = \|g \otimes g\|_p (= 1) = \|f \otimes f\|_p = \|T\|,$$

yet $X^*X - XX^* \neq 0$.

Example 4.6.8 Take $T = \begin{bmatrix} 3 & 0 \\ 0 & 3 \end{bmatrix} (\geq 0)$ and $X = \begin{bmatrix} 3 & \sqrt{6} \\ -3 & 3 \end{bmatrix}$ so that $T = T^*$ and $TX = XT$ (so the conditions of Theorem 4.6.3 are met). Then for $0 < p < 1$ we have the strict inequality

$$\|T - (X^*X - XX^*)\|_p^p = 6^p < 2.3^p = \|T\|_p^p.$$

(This example also shows that in the $p = 1$ case even if the conditions of Theorem 4.6.4 are met, there does not have to be a unique minimizer of $\|T-(X^*X- XX^*)\|_1$: for here

$$\|T - (X^*X - XX^*)\|_1(= 6) = \|T\|_1$$

yet $(X^*X - XX^*) \left(= \begin{bmatrix} 3 & 0 \\ 0 & -3 \end{bmatrix}\right) \neq 0).$

Exercises

1 Prove that, in finite dimensions, $\|I - (AX - XA)\| \geq \|I\|$ for all A and X.

2 Give an alternative proof Theorem 4.3.2(a) using the Kleinecke–Shirokov Theorem.

3 Let l_2 be the space of square-summable sequences of complex numbers and let S be the simple unilateral shift on l_2 given by $S(x_1, x_2, x_3, \ldots) = (0, x_1, x_2, \ldots)$. Verify that S is hyponormal.

4 Prove that if X^* is hyponormal and if T is an isometry such that $X^*T = TX^*$ then

$$\|T - (X^*X - XX^*)\| = \|T\|.$$

5 Construct a counter-example to show that Theorem 4.3.16 breaks down if $XT \neq TX$.

6 An operator A is said to be paranormal if

$$\|Af\|^2 \leq \|A^2f\|\|f\|$$

for all f in H. It can be shown that every hyponormal operator is paranormal. Prove that:

(a) if A is paranormal then $\|A^n\| = \|A\|^n$;
(b) if A is paranormal and if T is an isometry such that $AT = TA$ then

$$\|T - (AX - XA)\| \geq \|T\|;$$

(c) in Theorems 4.3.2(b), 4.3.16 and in question 4 above prove that "hyponormal" may be replaced by "paranormal".

7 (a) Let E be a linear space over $\mathbb{C}$ and A be in $L(E)$. Prove that Ran$A \cap$ Ker$A = \{0\}$ if and only if Ker$(A^n) =$ Ker$A$ for $n \geq 1$.

(b) Deduce that if the pair (A, B) has the property $(F\text{-}P)_{C_p}$, for $1 \leq p, \infty$, then, for $n \geq 1$,

$$\text{Ker}(\Delta_{A,B}^n \big| C_p) = \text{Ker}(\Delta_{A,B|C_p}).$$

8 Prove directly (i.e, without using Theorems 4.5.1 or 4.5.2) that, with A and B satisfying the conditions of Theorem 4.5.1, if $AV - VB = 0$ then V is a critical point of $X \to \|T - (AX - XB)\|_p^p$ for $1 < p < \infty$.

9 Prove that if the pair(A, B) satisfy the $(F\text{-}P)_{C_p}$ property, where $1 < p < \infty$, and if

$$A|W|^{p-1}U^* = |W|^{p-1}U^*B,$$

where $W = U|W|$ is the polar decomposition of W and where $W \in C_p$ for $1 < p < \infty$, then

$$A|W|U^* = |W|U^*B.$$

Notes

We make some further bibliographical comments in addition to those already made. Section 4.1 is adapted from part of [38, Chap. 5]. Theorem 4.2.4 is in [32, Lemma 3.1]. Anderson's ingenious proofs of the results Theorems 4.3.5 ... Theorems 4.3.7 are in [5, (1.4), (1.5), (1.6), (1.7)]. Anderson and Foias prove a more powerful version of Theorem 4.3.9 in [6, Theorem 1.5]. Theorem 4.3.14 is adapted from Bouali and Cherki [10, Theorem 2.2]. Theorem 4.3.16 is a special case of [36, Theorem 3.2]. Sections 4.4, 4.5 and 4.6 are based on [32], [10] and [35] respectively.

As for the exercises, Exercise 3 is in [36, Example 3.6]. Exercise 6(a), (b) are in [36, Theorem 2.3, Theorem 3.2].

Chapter 5
Spectral, and Numerical Range, Approximants

The theory of spectral approximants presents a precise geometric way of specifying approximants. The theory was initiated by Halmos [22] and later extended to the context of C_p by Bouldin [11] and Bhatia [8]. More recently the related concept of numerical range approximant was introduced [25].

5.1 Spectral Approximants in $\mathcal{L}(H)$

The idea of spectral approximant is this: a set E in the complex plane is specified corresponding to which there is a set $\mathcal{S}(E)$ consisting precisely of those operators each of whose spectrum is in E; then a **spectral approximant** of some given operator A is an approximant of A from $\mathcal{S}(E)$. For instance, the case when $E = \mathbb{R}$ corresponds to self-adjoint approximation and when $E = \mathbb{R}^+$ to positive approximation.

The concept of spectral approximant, as just stated, is too general for many positive results to be made. For instance, the set $\mathcal{S}(E)$ need not be closed. The standard example [23, Problem 104] is of a sequence of quasinilpotent operators (that is, of operators each of whose spectrum is $\{0\}$) whose limit is not quasinilpotent.

It is more fruitful to consider the set $\mathcal{J}(E)$ consisting precisely of those **normal** operators each of whose spectrum is in E:

$$\mathcal{J}(E) = \{X : X \text{ is normal and } \sigma(X) \subseteq E\}.$$

Yet even here there is an example, due to Rogers [41, Example 1.8] of a non-normal operator A and of a closed set E (viz. the unit circle) for which the distance $d(A, \mathcal{J}(E)) = \inf\{\|A - X\| : X \in \mathcal{J}(E)\}$ is not attained (it should be mentioned that if E is closed then $\mathcal{J}(E)$ is closed [22, p. 52]). Fortunately, under the assumptions

© Springer International Publishing AG 2017
P.J. Maher, *Operator Approximant Problems Arising from Quantum Theory*,
DOI 10.1007/978-3-319-61170-9_5

Fig. 5.1 Action of the
retraction F

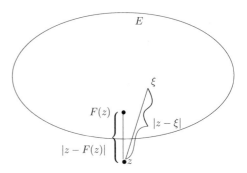

that

(i) E is a **closed** set in $\mathbb{C}$,
(ii) approximants are from $\mathcal{J}(E)$ and hence **normal**,
(iii) the operator A to be approximated to is **normal**

the theory is powerful and beautiful: under these assumptions an approximant of A from $\mathcal{J}(E)$ turns out to be a function of A (Fig. 5.1).

Definition 5.1.1 Let E be a non-empty subset of the complex plane $\mathbb{C}$. A (distance-minimizing) **retraction** for E is a function $F : \mathbb{C} \to E$ such that

$$|z - F(z)| \leq |z - \zeta|$$

for all ζ in E where $z \in \mathbb{C}$.

Observe that if $z \in E$ then $F(z) = z$ (hence the term "retraction") whilst the inequality justifies the term "distance-minimizing" (in the sequal the term "distance-minimizing" will be omitted).

Notice that a set E does not always uniquely determine a retraction onto E and that retractions need not be continuous: this is illustrated by the set $\{0, 1\}$ with retractions F_1 and F_2 given by

$$F_1(z) = \begin{cases} 1 & \text{if } \mathcal{R}z > \frac{1}{2} \\ 0 & \text{if } \mathcal{R}z \leq \frac{1}{2} \end{cases}, \qquad F_2(z) = \begin{cases} 1 & \text{if } \mathcal{R}z \geq \frac{1}{2} \\ 0 & \text{if } \mathcal{R}z < \frac{1}{2}. \end{cases}$$

Fortunately, retractions—especially those for convex sets—have some nice properties as is shown by Lemma 5.1.2 (proofs of the properties listed there and further comments are in [22, pp. 54–55]).

Lemma 5.1.2 *Let E be a non-empty, closed set in $\mathbb{C}$. Then:*

(a) there exists a Borel measurable retraction for E;
(b) the following are equivalent:

> *E is convex,*
> *E has no more than one retraction,*
> *E has a continuous retraction.*

We also require (here, and later in the context of $\mathcal{C}_p$) the following result. Throughout, let dist$(\alpha, S)(= \inf\{|\alpha-z| : z \in S\})$ denote the distance from a complex number α to some set S in $\mathbb{C}$.

Lemma 5.1.3 *Let X be normal, α be a complex number and f be a unit vector. Then*

$$\|(\alpha I - X)f\| \geq \operatorname{dist}(\alpha, \sigma(X));\tag{5.1.1}$$

and if there is a unique point β in $\sigma(X)$ which is closest to α and if equality holds in (5.1.1) then f is an eigenvector of X with corresponding eigenvalue β.

Proof Let $P(\cdot)$ be the spectral measure of X; then by (2.2.2)

$$\|(\alpha I - X)f\|^2 = \int_{\sigma(X)} |\alpha - z|^2 d\, \langle P(z)f, f\rangle$$

$$\geq \int_{\sigma(X)} \operatorname{dist}(\alpha, \sigma(X))^2 d\, \langle P(z)f, f\rangle$$

$$= \operatorname{dist}(\alpha, \sigma(X))^2$$

as desired.

Suppose that equality holds in (5.1.1) and β is the unique point in $\sigma(X)$ closest to α. Then $|\alpha - z| = \operatorname{dist}(\alpha, \sigma(X))$ so that $z = \beta$ almost everywhere with respect to the measure $\langle P(\cdot)f, f\rangle$. Therefore,

$$\|(\beta I - X)f\|^2 = \int_{\sigma(X)} |\beta - z|^2 d < P(z)f, f >= 0,$$

that is, $Xf = \beta f$.   □

Theorem 5.1.4 *Let E be a non-empty, closed set in the complex plane $\mathbb{C}$ with Borel measurable retraction F, let (as usual) $\mathcal{J}(E) = \{X : X \text{ is normal and } \sigma(X) \subseteq E\}$ and let A be normal. Then $F(A) \in \mathcal{J}(E)$ and*

$$\|A - F(A)\| \leq \|A - X\|\tag{5.1.2}$$

for all X in $\mathcal{J}(E)$.

Remark 5.1.5 Observe how beautifully the norm inequality (5.1.2) in this result replicates the modulus inequality that characterizes retractions.

Proof The proof hinges on the spectral theorem: represent A as a multiplication by ϕ on some L_2 space (Sect. 2.2). Since the retraction F is bounded on bounded sets [22, p. 54] the functional calculus makes sense; and as A is normal so is $F(A)$.

Since $\mathrm{Ran}F \subseteq E$ then $\mathrm{Ran}F \circ \phi \subseteq E$. Since E is closed then $\mathrm{essRan}F \circ \phi \subseteq E$ ("essRan" means essential range [23, Problem 123, p 67]); and since $\mathrm{essRan}F \circ \phi = \sigma(F(A))$ then $\sigma(F(A)) \subseteq E$. Thus $F(A) \in \mathcal{J}(E)$.

The proof of the inequality proceeds in a number of steps. Throughout, let $\rho : \mathbb{C} \to \mathbb{R}^+$ be the function given by $\rho(\alpha) = \mathrm{dist}(\alpha, E)$.

(i) *If α is an eigenvalue of A then $\rho(\alpha) \leq \|A - X\|$.*
Proof of (i). Let f be a unit vector corresponding to the eigenvalue α. Then, by Lemma 5.1.3,

$$\rho(\alpha) \leq \mathrm{dist}(\alpha, \sigma(X)) \leq \|(\alpha I - X)f\| = \|(A - X)f\| \leq \|A - X\|.$$

(ii) *If $\alpha \in \sigma(A)$ then there exists a sequence $\{A_k\}$ of normal operators each having α as an eigenvalue and such that $\|A_k - A\| \to 0$ as $k \to \infty$.*
Proof of (ii). $\alpha \in \sigma(A)$ means $\alpha \in \mathrm{essRan}\phi$. To construct the sequence $\{A_k\}$ let D_k be the open disc with centre α and diameter $\dfrac{1}{k}$; as $\alpha \in \mathrm{essRan}\phi$ then $\phi^{-1}(D_k)$ has positive measure; let $\phi_k = \alpha$ in $\phi^{-1}(D_k)$ and $\phi_k = \phi$ outside $\phi^{-1}(D_k)$ and let A_k be the operator corresponding to multiplication by ϕ_k. Then each A_k is normal and has α as an eigenvalue. As $|\phi_k - \phi| < \frac{1}{k}$ inside $\phi^{-1}(D_k)$ and $|\phi_k - \phi| = 0$ outside $\phi^{-1}(D_k)$ then $\phi_k \to \phi$ uniformly and hence $A_k \to A$ in norm as $k \to \infty$.

(iii) *If $\alpha \in \sigma(A)$ then $\rho(\alpha) \leq \|A - X\|$.*
Proof of (iii). By (ii) there is a sequence $\{A_k\}$ of normal operators, each having α as an eigenvalue, such that $\|A_k - A\| \to 0$ as $k \to \infty$. By (i), $\rho(\alpha) \leq \|A_k - X\|$ for every k so that $\rho(\alpha) \leq \|A - X\|$

(iv) $\|A - F(A)\| \leq \|A - X\|$.
Proof of (iv). The normal operator $A - F(A)$ is represented by multiplication by $\phi - F \circ \phi$. Since

$$|\phi - F \circ \phi| = \rho \circ \phi$$

and since almost every value of ϕ is in $\sigma(A)$ it therefore follows from (iii) that

$$|\phi - F \circ \phi| \leq \|A - X\|$$

almost everywhere. The inequality (5.1.2) follows immediately.

$\square$

The proof of the inequality (5.1.2) for convex E is somewhat crisper than the general case.

Proof of Inequality (5.1.2) *of Theorem 5.1.4 for convex E.* First, suppose A has an eigenvalue α with corresponding eigenvector f of unit norm so that $\alpha = \langle Af, f \rangle$. Then, as F is a retraction onto E,

$$\|(A - F(A))f\| = |\alpha - F(\alpha)| \le |\alpha - \zeta|$$

for all ζ in E; in particular, we may take $\zeta = \langle Xf, f \rangle$ since, with W denoting numerical range, we have, using (2.1.1),

$$\langle Xf, f \rangle \in W(X) \subseteq \overline{W(X)} = \text{conv}\,\sigma(X) \subseteq \text{conv}E = E \tag{5.1.3}$$

by the convexity of E. Thus,

$$\|(A - F(A))f\| \le |\langle (A - X)f, f \rangle| \le w(A - X) \le \|A - X\|$$

where w is the numerical radius (See Sect. 2.1). The inequality (5.1.2) is immediate on taking the supremum norm over f.

If A is diagonal (i.e. has an orthonormal basis consisting of eigenvectors of A) then the inequality (5.1.2) follows from the above conclusion.

If, finally, A is an arbitrary normal operator it is, by the spectral theorem (Sect. 2.2), the norm limit of diagonal ones. Since E is convex the (unique) retraction F is continuous; and hence the mapping $X \to F(X)$, for normal operators X, is continuous. The inequality (5.1.2) is now immediate from the previous conclusion.
□

5.2   Spectral Approximants in C_p

Spectral approximation in C_p differs from the $\mathcal{L}(H)$ case in two ways. First, in minimizing the quantity

$$\|A - X\|_p$$

we have to make the obvious assumption that

$$X \in \mathcal{J}(E) \quad \text{and} \quad A - X \in C_p.$$

Theorem 5.2.3 states a necessary and sufficient condition for there to exist such an X.

Second, as is justified by Lemma 5.2.2, there is the (somewhat irritating) restriction on the parameter p that $2 \leq p < \infty$ if the set E is (as usual) non-empty and closed (as distinct from being, additionally, convex, or balanced: see Theorem 5.2.6).

Bouldin's first step [11, Lemma 1] towards his result (Theorem 5.2.4) on normal spectral approximants in $\mathcal{C}_p$ is the result below.

Lemma 5.2.1 *Let A be normal. If there exists some X in $\mathcal{J}(E)$ such that $A - X \in \mathcal{C}_p$ where $1 \leq p < \infty$ then the only points in $\sigma(A)$ not contained in E are isolated eigenvalues of A with finite multiplicity.*

Proof The hypothesis that $A - X \in \mathcal{C}_p$ says that A is a compact perturbation of X, that is, $A = X + K$ for compact K. From Weyl's Theorem for normal operators, A and X must have the same Weyl spectrum. Now, for some normal operator T, say, the Weyl spectrum of T coincides with the points of $\sigma(T)$ that are not isolated eigenvalues with finite multiplicity [7, Theorem 5.1]. Since the Weyl spectrum of X, and hence the Weyl spectrum of A, is contained in E (because $X \in \mathcal{J}(E)$) it follows that the only points in $\sigma(A)$ **not** contained in E **are** the isolated eigenvalues of A with finite multiplicity (For an exposition of the Weyl spectrum see [7]) (Fig. 5.2).   □

Lemma 5.2.2 *Let $\{\phi_i\}$ be a complete orthonormal sequence and T be in $\mathcal{C}_p$, where $2 \leq p < \infty$. Then*

$$\|T\|_p^p \geq \sum_{i=1}^{\infty} \|T\phi_i\|^p.$$

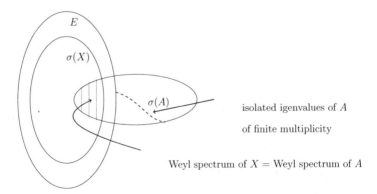

Fig. 5.2 The Weyl spectrum

Proof Using (2.4.7) we have

$$\|T\|_p^p = \||T|\|_p^p = \sum_{i=1}^{\infty} s_i(|T|)^p$$

$$= \sum_{i=1}^{\infty} s_i(|T|^2)^{p/2} = \||T|^2\|_{p/2}^{p/2}$$

$$\geq \sum_{i=1}^{\infty} \langle |T|^2\phi_i, \phi_i\rangle^{p/2} \quad \text{provided} \quad \frac{p}{2} \geq 1$$

$$= \sum_{i=1}^{\infty} \|T\phi_i\|^p.$$

□

Observe that when $p = 2$ we automatically have the equality $\|T\|_2^2 = \sum \|T\phi_i\|^2$ and for $1 \leq p < 2$ the reverse inequality may be strict: see Exercise 1.

Theorem 5.2.3 *Let A be normal. In order for there to exist some X in $\mathcal{J}(E)$ such that $A - X \in \mathcal{C}_p$, where $2 \leq p < \infty$, it is necessary and sufficient that $\sigma(A)/E$ is a possible empty, finite or infinite countable set $\{\alpha_i\}$ of isolated eigenvalues of finite multiplicity, repeated according to multiplicity, such that $\sum_i (\mathrm{dist}(\alpha_i, E))^p$ is finite.*

Proof See Exercise 5.   □

For an operator A and a non-empty, closed set E we define the set $\mathcal{J}_p(E)$ as follows:

$$\mathcal{J}_p(E) = \{X : A - X \in \mathcal{C}_p \text{ and } X \in \mathcal{J}(E)\}.$$

Theorem 5.2.4 *Let A be normal, F be a Borel measurable retraction for the non-empty, closed set E and let X vary such that $X \in \mathcal{J}_p(E)$ for $2 \leq p < \infty$. Then $F(A) \in \mathcal{J}_p(E)$ and*

$$\|A - F(A)\|_p \leq \|A - X\|_p; \tag{5.2.1}$$

further, $F(A)$ is the unique choice of X producing equality in (5.2.1) if and only if every point of $\sigma(A)$ has a unique closest point in E; and, in particular, if E is convex then equality in (5.2.1) implies $X = F(A)$.

Proof As $A - X \in \mathcal{C}_p$ then, in accordance with Lemma 5.2.1, let $\{\phi_i\}$, where $1 \leq i \leq l \leq \infty$, be a maximal orthonormal set of eigenvectors of A corresponding to the (countable set of) isolated eigenvalues $\{\alpha_i\}$ of A not contained in E (where the α_i are in decreasing order of (finite) multiplicity). By Lemmas 5.2.2 and 5.1.3 we have, provided $2 \leq p < \infty$,

$$
\begin{aligned}
\|A - X\|_p^p &\geq \sum_{i=1}^{l} \|(A - X)\phi_i\|^p \\
&= \sum_{i=1}^{l} \|(\alpha_i I - X)\phi_i\|^p \\
&\geq \sum_{i=1}^{l} \text{dist}(\alpha_i, \sigma(X))^p \\
&\geq \sum_{i=1}^{l} \text{dist}(\alpha_i, E)^p \\
&= \sum_{i=1}^{l} |\alpha_i - F(\alpha_i)|^p
\end{aligned}
\tag{5.2.2}
$$

where the last inequality follows because $\sigma(X) \subseteq E$ and the last equality follows because F is a retraction onto E.

We now prove that this last sum is $\|A - F(A)\|_p^p$. Express $A = A_1 \oplus A_2$ relative to the decomposition

$$H = P(E)H \oplus P(E^c)H$$

where $P(\cdot)$ is the spectral measure of A and E^c means the complement of E. Observe that

$$A_1 \in \mathcal{J}(E) \text{ and } A_2 = \sum_{i=1}^{l} \alpha_i(\phi_i \otimes \phi_i)$$

and so, since $F(z) = z$ for all z in E,

$$F(A) = F(A_1) \oplus F(A_2) = A_1 \oplus \sum F(\alpha_i)(\phi_i \otimes \phi_i) \in \mathcal{J}(E).$$

(The point here is not that $F(A) \in \mathcal{J}(E)$, which has already been proved in Theorem 5.1.4, but rather that $F(A_1)$ and $F(A_2)$ are each in $\mathcal{J}(E)$.) Let $\{\psi_j\}$ be an orthonormal basis of $P(E)H$; then $\{\phi_1, \ldots, \phi_l, \psi_1, \ldots\}$ diagonalizes $A - F(A)$ with corresponding eigenvalues $\{\alpha_1 - F(\alpha_1), \alpha_2 - F(\alpha_2), \ldots, \alpha_l - F(\alpha_l), 0, 0, \ldots, 0\}$. Hence,

$$\sum_{i=1}^{l} |\alpha_i - F(\alpha_i)|^p = \|A - F(A)\|_p^p$$

as desired (observe the argument in this paragraph is valid for $1 \leq p < \infty$).

As for uniqueness, suppose on the one hand, that $\sigma(A)$ has a unique closest point in E; let X be an operator in $\mathcal{J}(E)$ for which equality holds in (5.2.1). Then equality holds throughout (5.2.2) which, in view of Lemma 5.1.3 (on putting f of Lemma 5.1.3 equal to ϕ_i for $i = 1, \ldots, l$) forces

$$\|(\alpha_i - X)\phi_i\| = \text{dist}(\alpha_i, \sigma(X)) = \text{dist}(\alpha_i, E).$$

From Lemma 5.1.3 it follows that ϕ_i is an eigenvector of X with corresponding eigenvalue $F(\alpha_i)$ (for $F(\alpha_i)$ is the (unique) point in E closest to α_i as F is a retraction onto E). Choosing $\{\psi_i\}$ as above, it follows from Lemma 5.2.2 that, for $2 \le p < \infty$,

$$\|A - X\|_p^p \ge \sum_{i=1}^{l} \|(A - X)\phi_i\|^p + \sum_{j} \|(A - X)\psi_j\|^p.$$

But since equality holds throughout (5.2.2) therefore

$$\|(A - X)\psi_j\| = 0 \qquad \text{for } j = 1, 2, \ldots.$$

So: since $\{\psi_j\}$ is a basis of $P(E)H$ then the restrictions to $P(E)H$ of X, A and hence of $F(A)$ coincide; and since X and $F(A)$ coincide on $S\{\phi_i\} = P(E^c)H$ (because, by Lemma 5.1.3, $X\phi_i = F(\alpha_i)\phi_i = F(A)\phi_i$), we conclude that $X = F(A)$ everywhere.

In the special case where E is convex every point in the complex plane has a unique nearest point in E so the proceeding argument yields $X = F(A)$.

On the other hand, suppose that $\sigma(A)$ does not have a unique closest point in E; that is, suppose there is some λ in $\sigma(A)$ for which

$$|\lambda - \mu| = |\lambda - F(\lambda)| \quad \text{where } \mu \ne F(\lambda) \in E \text{ and } \mu \in E.$$

Then the definition of F can be altered by setting $F(\lambda) = \mu$ so that there are two distinct Borel measurable retractions onto E which differ on $\sigma(A)$: thus, $F(A)$ is not the unique best approximant.   □

The Bouldin $\| \cdot \|_p$ inequality of Theorem 5.2.4 does not necessarily hold for $1 \le p < 2$ for non-convex E.

Example 5.2.5 ([8, p. 35]) Let $H = \mathbb{C}^2$ and E be the closed set $E = \{i, -i\}$ (note: E is not convex). Let A and X be given by

$$A = \begin{bmatrix} 0 & 1 \\ 1 & 0 \end{bmatrix}, \quad X = \begin{bmatrix} 0 & 1 \\ -1 & 0 \end{bmatrix}.$$

Then $X \in \mathcal{J}(E)$ and $\|A - X\|_p = 2$. Let F be any function on the plane onto E. Then since $F(A)$ and A commute they can be simultaneously diagonalized and it can be checked that

$$\|A - F(A)\|_p = 2^{\frac{1}{p} + \frac{1}{2}}.$$

Hence, for $1 \le p < 2$, $\|A - F(A)\|_p > \|A - X\|_p$.

The Bouldin $\|\cdot\|_p$ inequality of Theorem 5.2.4 can be extended to the $1 \le p < \infty$ case provided the set E is convex or is balanced. Recall: a set E in $\mathbb{C}$ is **balanced** if $z \in E \implies e^{i\theta} z \in E$ for all θ.

Theorem 5.2.6 *Let A be normal, F be a Borel measurable retraction for the non-empty, closed set E and let X vary such that $X \in \mathcal{J}_p(E)$ for $1 \le p < \infty$. Then:*

(a) *if E is convex it follows that $F(A) \in \mathcal{J}_p(E)$ and*

$$\|A - F(A)\|_p \le \|A - X\|_p;$$

(b) *if E is assumed balanced, rather than convex, the same conclusion as in (a) holds.*

Proof

(a) Let $\{\phi_i\}$ and $\{\alpha_i\}$, where $1 \le i \le l \le \infty$, be as in Theorem 5.2.4 so that $\{\phi_i\}$ is a maximal orthonormal set of eigenvectors of A corresponding to the isolated eigenvalues $\{\alpha_i\}$ of A, not contained in E. Then, by Theorem 5.2.4, for $1 \le p < \infty$

$$\|A - F(A)\|_p^p = \sum_{i=1}^{l} |\alpha_i - F(\alpha_i)|^p$$

$$\le \sum_{i=1}^{l} |\alpha_i - \langle X\phi_i, \phi_i \rangle|^p$$

$$= \sum_{i=1}^{l} |\langle (A - X)\phi_i, \phi_i \rangle|^p$$

$$\le \|A - X\|_p^p$$

where the last inequality follows from (2.4.7) and the first inequality follows from the convexity of E, for exactly as in the convexity trick (5.1.3) (put $\phi = $ "f" there) we have $\langle X\phi_i, \phi_i \rangle \in E$ whence $|\alpha_i - F(\alpha_i)| \le |\alpha_i - \langle X\phi_i, \phi_i \rangle|$, for $1 \le i \le l$, since F is a retraction onto E.

(b) Again, as the starting point we have

$$\|A - F(A)\|_p^p = \sum_{i=1}^{l} |\alpha_i - F(\alpha_i)|^p. \tag{5.2.3}$$

Let $\alpha = |\alpha|e^{i\theta}$. Since F maps $\mathbb{C}$ onto the balanced set E then $F(\alpha) = |F(\alpha)|e^{i\theta}$ (for the same θ), and since F is a retraction on $\mathbb{R}^+$ onto $|E| = \{|\xi| : \xi \in E\}$ then $F(|\alpha|) = |F(\alpha)|$. Thus, for all α in $\mathbb{C}$

$$|\alpha - F(\alpha)| = ||\alpha| - |F(\alpha)|| = ||\alpha| - (F|\alpha|)| \le ||\alpha| - |\xi||$$

for all $|\xi|$ in $|E|$. Since A is normal then $|\alpha_i| = s_i(A)$ (the ith singular value of A) for $1 \le i \le l \le \infty$. Hence,

$$|\alpha_i - F(\alpha_i)| \le |s_i(A) - |\xi||.$$

Since $\sigma(X) \subseteq E$ it follows that the point spectrum $\sigma_p(X) \subseteq E$ and hence, as X is normal, each $s_j(X) \in |E|$. Hence, the $|\xi|$ occurring above may vary over all the singular values $s_j(X)$ for $1 \le j < \infty$ (the $s_j(X)$ being in decreasing order and repeated according to multiplicity). Hence, in particular,

$$|\alpha_i - F(\alpha_i)| \le |s_i(A) - s_i(X)|$$

for $1 \le i \le l \le \infty$. Therefore, from (5.2.3)

$$\|A - F(A)\|_p^p \le \sum_{i=1}^{l} |s_i(A) - s_i(X)|^p. \qquad (5.2.4)$$

From [43, 1.22] it follows that if $\displaystyle\sum_{j=1}^{\infty} s_i(A - X)^p (= \|A - X\|_p^p) < \infty$ then

$$\sum_{j=1}^{\infty} |s_i(A) - s_i(X)|^p \le \sum_{j=1}^{\infty} s_i(A - X)^p.$$

Therefore, from (5.2.4) $\|A - F(A)\|_p^p \le \|A - X\|_p^p.$

$\square$

The self-adjoint approximant result, Theorem 3.2.1, fits into the pattern of the spectral approximation results, Theorems 5.1.4 and 5.2.6(a): take $E = \mathbb{R}$ and $F(z) = \frac{z+\bar{z}}{2}$. But the spectral approximation results do not contain Theorem 3.2.1 (whose proof is so simple) since the operator to be approximated to in Theorem 3.2.1 is not necessarily normal.

Where the spectral approximant results come into their own is in positive approximation of a normal operator. Corollary 5.2.7 applies to both $\mathcal{L}(H)$ and $\mathcal{C}_p$; and its proof, compared to that of Theorem 3.3.3 (which, anyway, applies only to $\mathcal{L}(H)$), is startlingly simple. In Corollary 5.2.7 the symbols $\| \cdot \|_\infty$ mean the supremum norm on $\mathcal{L}(H)$.

Corollary 5.2.7 *Let A be normal and X vary over the positive operators and further be such that* $A - X \in \mathcal{C}_p$ *if* $1 \leq p < \infty$. *Then* $A - (\mathcal{R}A)^+ \in \mathcal{C}_p$ *for* $1 \leq p < \infty$ *and*

$$\|A - (\mathcal{R}A)^+\|_p \leq \|A - X\|_p, \qquad 1 \leq p \leq \infty, \qquad (5.2.5)$$

with equality occurring in (5.2.5) if, and for $1 \leq p < \infty$ *only if,* $X = (\mathcal{R}A)^+$.

Proof Take $E = \mathbb{R}^+$ and

$$F(z) = \begin{cases} \mathcal{R}z & \text{if } \mathcal{R}z \geq 0 \\ 0 & \text{if } \mathcal{R}z < 0 \end{cases},$$

so that F is a retraction onto E. Then $F(A) = (\mathcal{R}A)^+$ and from Theorem 5.1.4, for $p = \infty$, the inequality (5.2.5) follows; and from Theorem 5.2.6(a), for $1 \leq p < \infty$, the inclusion $A - (\mathcal{R}A)^+ \in \mathcal{C}_p$ and the inequality (5.2.5) follow since E is convex. The equality assertion follows from Theorem 2.4.1 since the set of positive operators is convex.   □

A special case of positive approximation (of some normal operator) is projection approximation. The next result shows that if there is a projection P such that $A - P \in \mathcal{C}_p$, for normal A, there is a projection that minimizes $\|A - P\|_p$ provided $2 \leq p < \infty$.

Corollary 5.2.8 *Let A be normal and P vary over the projections and be such that* $A - P \in \mathcal{C}_p$ *for* $2 \leq p < \infty$. *Then there exists a projection V such that* $A - V \in \mathcal{C}_p$ *for* $2 \leq p < \infty$, *and*

$$\|A - V\|_p \leq \|A - P\|_p, \qquad 2 \leq p \leq \infty.$$

Proof Take $E = \{(0,0), (1,0)\}$ so that $\sigma(P) = E$. Let F be a retraction onto E. Then $\sigma(F(A)) = E$ so that $F(A)$ is a projection, say V. The result follows from Theorems 5.1.4 and 5.2.4 (for $p = \infty$ and $2 \leq p < \infty$ respectively).   □

Unfortunately, Theorem 5.2.6 does not apply to projection approximation: the set $E = \{(0,0), (1,0)\}$ is neither convex nor balanced. In Chap. 6 we will extend Corollary 5.2.8 to the $1 \leq p < \infty$ case—as a spin off, there, from the study of partially isometric approximation.

5.3   Numerical Range Approximants

A numerical range approximant is defined similarly to a spectral approximant. For a given set E in $\mathbb{C}$, let $\mathcal{W}(E)$ be the set of all those operators each of whose numerical range is in E; then a **numerical range approximant** (with respect to the some norm) of some operator A is an approximant of A from $\mathcal{W}(E)$.

Though defined similarly to spectral approximants, numerical range approximants are better behaved: as we shall see in Sect. 5.4, every operator in $\mathcal{L}(H)$

has a numerical range approximant. Thus, there is not the same necessity (as in the spectral case) to restrict attention to **normal** numerical range approximants of **normal** operators. Nevertheless, under these assumptions we do have the analogue (Theorem 5.3.1 below) of the spectral approximation results (Theorems 5.1.4, 5.2.4 and 5.2.6). Let

$$\mathfrak{W}(E) = \{X : X \text{ is normal and } W(X) \subseteq E\}.$$

Theorem 5.3.1 *Let E be a non-empty, closed convex set in the complex plane with Borel measurable retraction F, let A be normal and let X vary such that $X \in \mathfrak{W}(E)$ and $A - X \in C_p$ if $1 \leq p < \infty$. Then $F(A) \in \mathfrak{W}(E), A - F(A) \in C_p$ if $1 \leq p < \infty$ and*

$$\|A - F(A)\|_p \leq \|A - X\|_p, \quad 1 \leq p \leq \infty. \tag{5.3.1}$$

Proof From Theorem 5.1.4 it follows that $F(A)$ is normal and $\sigma(F(A)) \subseteq E$. Hence, as E is convex, using the same convexity trick as in (5.1.3) we find that $W(F(A)) \subseteq E$ so that $F(A) \in \mathfrak{W}(E)$. Since each normal operator X is constrained to satisfy $W(X) \subseteq E$ and hence $\overline{W(X)} \subseteq \overline{E} = E$, that is, by (2.1.1), $\text{conv}\sigma(X) \subseteq E$, it follows that $\sigma(X) \subseteq E$. Hence, for $p = \infty$ the inequality (5.3.1) follows from Theorem 5.1.4; and for $1 \leq p < \infty$ the inequality (5.3.1) and the inclusion $A - F(A) \in C_p$, follow from Theorem 5.2.6(a) since E is convex. □

5.4 Proximality

A concept that is pertinent to the whole circle of ideas explored in this work is that of proximality. A subset $\mathcal{K}$ of some Banach space $\mathcal{L}$ in $\mathcal{L}(H)$ is said to be **proximal in $\mathcal{L}$** if to each A in $\mathcal{L}$ there corresponds an X_1, say, in $\mathcal{K}$ which minimizes $\|A - X\|$ where $X \in \mathcal{K}$. For a unitarily invariant norm $\|\cdot\|$, with associated ideal $\mathcal{C}$, a subset $\mathcal{K}$ of some Banach space $\mathcal{L}$ in $\mathcal{L}(H)$ is said to be **proximal in $\mathcal{L}$ with respect to** $\|\cdot\|$ if to each A in $\mathcal{L}$ the set

$$\{X : X \in \mathcal{K} \text{ and } A - X \in \mathcal{C}\} = \mathcal{S}, \text{ say,}$$

is non-empty and if there corresponds an X_1 in $\mathcal{S}$ which minimizes $\|A - X\|$ for all X in $\mathcal{S}$.

Example 5.4.1

(a) The set of all self-adjoint operators is proximal in $\mathcal{L}(H)$ with respect to all unitarily invariant norms (Theorem 3.2.1).
(b) The set of positive operators is proximal in $\mathcal{L}(H)$ (Theorem 3.3.1); and is proximal in the set $\mathcal{N}$ normal operators with respect to $\|\cdot\|_p$ for $1 \leq p < \infty$ (Corollary 5.2.7).

(c) The set of unitary operators is not proximal in $\mathcal{L}(H)$ (This is Rogers' example [41, Example 1.8] cited in Sect. 5.1).
(d) For a non-empty, closed set E, the set $\mathcal{J}(E)$ $[\mathcal{J}_p(E)]$ of operators is proximal in $\mathcal{N}$ [with respect to $\|\cdot\|_p$ for $2 \le p < \infty$] (Theorems 5.1.4 and 5.2.4); and if, additionally, E is assumed convex (or balanced) $\mathcal{J}_p(E)$ is proximal in $\mathcal{N}$ with respect to $\|\cdot\|_p$ for $1 \le p < \infty$ (Theorem 5.2.6).
(e) For a non-empty, closed, convex set E the set $\mathfrak{W}(E)$ of operators is proximal in $\mathcal{N}$ with respect to $\|\cdot\|_p$ for $1 \le p \le \infty$ (Theorem 5.3.1).

The result (d) that $\mathcal{J}(E)$ is proximal in $\mathcal{N}$ is a consequence of Theorem 5.1.4. In fact, the proximality of $\mathcal{J}(E)$ in $\mathcal{N}$ can be proved from "first principles" without recourse of Theorem 5.1.4. The reader is invited to do this in Exercise 7. We also have the following "first principles" result.

Theorem 5.4.2 *Let E be a non-empty, closed set in the complex plane; then $\mathfrak{W}(E)$ is proximal in $\mathcal{L}(H)$.*

Remark 5.4.3 This illustrates that existence of a relevant approximant—proximality—is weaker than its identification. Theorem 5.4.2 says that every operator in $\mathcal{L}(H)$ has a numerical range approximant but does not identify it.

Proof Let A be in $\mathcal{L}(H)$ and let $d(A, \mathfrak{W}(E)) = r$ so that there exists a sequence $\{B_n\}$ in $\mathfrak{W}(E)$ such that $\lim_{n \to \infty} \|A - B_n\| = r$. The sequence $\{B_n\}$ is bounded (since $\|B_n\| \le \|B_n - A\| + \|A\| \le r + \|A\|$). As $\mathcal{L}(H)$ is the dual of $\mathcal{C}_1$ (Sect. 2.4) and as $\mathcal{C}_1$ is separable, the sequence $\{B_n\}$ has a subsequence $\{B_{n_k}\}$ which converges to some bounded operator B, say, in the weak operator topology; that is, $\lim_{n_k \to \infty} \langle B_{n_k} f, g \rangle = \langle Bf, g \rangle$ for all f and g in H. In particular, $\lim_{n_k \to \infty} \langle B_{n_k} f, f \rangle = \langle Bf, f \rangle$ where $f \in H$ and $\|f\| = 1$. As $\langle B_{n_k} f, f \rangle \in E$ for all n and all f such that $\|f\| = 1$, it follows, that since E is closed, $W(B) \subseteq E$, that is, $B \in \mathfrak{W}(E)$.

To see that B is an approximant of A note that if $f \in H, g \in H$ with $\|f\| = \|g\| = 1$ then

$$| \langle (A - B)f, g \rangle | = \lim_{n_k \to \infty} | \langle (A - B_{n_k})f, g \rangle |$$

$$\le \lim_{n_k \to \infty} \|A - B_{n_k}\| \|f\| \|g\| = r$$

so that

$$\|A - B\| = \sup | \langle (A - B)f, g \rangle | = r (= \inf_{X \in \mathfrak{W}(B)} \|A - X\|)$$

as desired.   □

A technique similar to that used in the proof of Theorem 5.4.2 may be used to solve Exercise 4 which is generalization of Theorem 3.3.1.

Exercises

1 Let $T = \begin{bmatrix} 2 & 1 \\ 1 & 0 \end{bmatrix}$. Show that $\|T\|_1 < \sum_{i=1}^{\infty} \|T\phi_i\|$.

2 Let $H = \mathbb{C}^3$. For A, P and Q given by

$$A = \begin{bmatrix} 0 & 0 & 0 \\ 0 & 1 & 0 \\ 0 & 0 & 2 \end{bmatrix}, \quad P = \begin{bmatrix} 0 & 0 & 0 \\ 0 & 1 & 0 \\ 0 & 0 & 1 \end{bmatrix}, \quad Q = \begin{bmatrix} \frac{1}{2} & \frac{1}{2} & 0 \\ \frac{1}{2} & \frac{1}{2} & 0 \\ 0 & 0 & 1 \end{bmatrix}$$

verify that for (the projections) P and Q

$$\|A - P\| = \|A - Q\| \leq \|A - X\|$$

for all projections X in $\mathcal{L}(H)$. How does Q differ from P?

3 Let $H = \mathbb{C}^2$ and $E = \{0\}$ (so that in dealing with $\mathcal{J}(E)$ we are considering nilpotent approximation). Let

$$A = \begin{bmatrix} 2 & 0 \\ 0 & 0 \end{bmatrix}.$$

Verify, by means of examples, that there is strict inequality:

$$\text{dist}(A, \mathcal{S}(E)) < \text{dist}(A, \mathcal{J}(E)).$$

(Recall: $\mathcal{S}(E)$ refers to spectral approximation without any normality requirement).

4 Let $E : \{(x, y) : 0 \leq x \leq \infty, 0 \leq y \leq a \leq \infty\}$. For A in $\mathcal{L}(H)$ let $A_1 = \mathcal{R}A$ and $A_2 = \mathcal{I}mA$. Let $\mathcal{D}(A)$ and $\mathcal{H}(A)$ be given by

$$\mathcal{D}(A) = \{\|A - X\| : X \in \mathcal{S}(E)\}$$

$$\mathcal{H}(A) = \{r : A_1 + \sqrt{r^2 - (A_2 - B)^2} \geq 0 \text{ for some } B \text{ in } \mathcal{S}([0, a])\}.$$

Prove that:

(a) $\mathcal{S}(E)$ is proximal in $\mathcal{L}(H)$;
(b) $\mathcal{D}(A) = \mathcal{H}(A)$;
(c) if $\hat{A} = A_1 + \sqrt{r^2 - (A_2 - B)^2} + i\hat{B}$, where $\hat{B}$ is a certain operator in $\mathcal{S}([0, a])$ and $r = \inf \mathcal{D}(A)$, then $\hat{A} \in \mathcal{S}(E)$ and, for all for all X in $\mathcal{S}(E)$

$$\|A - \hat{A}\| \leq \|A - X\|.$$

5 Prove Theorem 5.2.3

6 An alternative approach to (and generalization of) Theorems 5.1.4, 5.2.4, 5.2.6(a). This question asks the reader to obtain Bhatia's extension of Theorems 5.1.4, 5.2.4, 5.2.6(a) to all so-called Q norms and to all unitarily invariant norms provided E is convex.

A norm is called a **Q norm**, denoted by $\| \cdot \|_Q$, if there exists a unitarily invariant norm $\|| \cdot \||'$ such that $\|A\|_Q = (\||A^*A\||')^{\frac{1}{2}}$. Examples: the sup norm $\| \cdot \|$ is a Q norm since $\|A^*A\| = \|A\|^2$, so that $\|A\| = (\|A^*A\|)^{\frac{1}{2}}$, the Von Neumann-Schatten norms $\| \cdot \|_p$ are Q norms provided $p \geq 2$ since, by (2.4.3), $\|A^*A\|_{\frac{p}{2}} = (\tau |A^*A|^{\frac{p}{2}})^{\frac{2}{p}} = (\tau |A|^p)^{\frac{2}{p}} = \|A\|_p^2$ so that $\|A\|_p = (\|A^*A\|_{\frac{p}{2}})^{\frac{1}{2}}$. Obviously, Q norms are unitarily invariant.

An inequality concerning unitarily invariant norms (and hence, in particular, Q norms) involves the Ky Fan norm. For a compact operator A, with singular values $s_1(A) \geq \ldots \geq s_k(A) \geq \ldots$ (repeated according to multiplicity) the **Ky Fan norms** $\| \cdot \|_K, K = 1, 2, \ldots$ are defined by $\|A\|_K = \sum_{i=1}^{K} s_i(A)$. The Ky Fan norms are unitarily invariant and satisfy the **Ky Fan Dominance Property:** if B belongs to the norm ideal $\mathcal{I}$ generated by some unitarily invariant norm $\|| \cdot \||$ and if $\|A\|_K \leq \|B\|_K$ for $K = 1, 2, \ldots$ then $A \in \mathcal{I}$ and $\||A\|| \leq \||B\||$.

Armed with this prove the following theorem.

Theorem 5.5.1 *Let E be a non-empty closed set with Borel measurable retraction F and let A be normal.*

(a) *Let $\| \cdot \|_Q$ be a Q norm and $\mathcal{C}$ its associated ideal. If there exists some X in $\mathcal{J}(E)$ such that $A - X \in \mathcal{C}$ then $A - F(A) \in \mathcal{C}$ and*

$$\|A - F(A)\|_Q \leq \|A - X\|_Q;$$

(b) *if, further, the set E is convex then under the hypothesis of (a)*

$$\||A - F(A)\|| \leq \||A - X\||$$

for an arbitrary unitarily invariant norm $\|| \cdot \||$.

Now deduce Theorem 5.2.6(a).

7 Prove from "first principles" (cf. proof of Theorem 5.4.2) that $\mathcal{J}(E)$ is proximal in $\mathcal{N}$.

8 Adopt the notation $\mathcal{W}_p(E) = \{X : A - X \in \mathcal{C}_p \text{ and } X \in \mathcal{W}(E)\}$. Let $E = E_1 \times E_2 \simeq \{x_1 + ix_2 : x_1 \in E_1 \text{ and } x_2 \in E_2\}$ where E_1 and E_2 are closed intervals; let $P_i(x_1, x_2) = x_i$ and $F_i : \mathbb{C} \to E_i$ be the retraction onto E_i for $i = 1, 2$; and let x be in $\mathcal{W}(E)$ [be in $\mathcal{W}_p(E)$ for $1 \leq p < \infty$ provided $\mathcal{W}_p(E) \neq \emptyset$].

(a) If $P_1(\mathcal{W}(A)) \subseteq E_1$ then $A_1 + iF_2(A_2) \in \mathcal{W}(E)$ [$\in \mathcal{W}_p(E)$ for $1 \leq p < \infty$] and for all X in $\mathcal{W}(E)$ [$\in \mathcal{W}_p(E)$]

$$\|A - (A_1 + iF_2(A_2))\|_p \leq \|A - X\|_p$$

for $1 \leq p < \infty$ with equality if, and for $1 < p < \infty$ only if, $X = A_1 + iF_2(A_2)$.

(b) Formulate, and prove a corresponding result beginning with the hypothesis "$P_2(\mathcal{W}(A)) \subseteq E_2$".

Notes

The key result, Theorem 5.1.4, on spectral approximation is due to Halmos [22, Theorem]. Halmos' work was extended to C_p for $p \geq 2$ by Bouldin: Lemma 5.2.1 and Theorems 5.2.3 and 5.2.4 are due to Bouldin [11, Lemma 1 and Theorems 1 and 2]. Bhatia's extension of Bouldin's result to the $1 \leq p < \infty$ case for convex E is in [8, Theorem 1]. Maher, Theorem 5.2.6 [34, Theorem 1] gives an alternative proof of Bhatia's result just cited and proves a similar result for balanced sets. The paper [25] also introduced the concept of proximality, as well as that of a numerical range approximant.

Exercises: Exercise 2 is due to Halmos [22, p. 53]; Exercise 4 is in Khalil and Maher [25, Theorem 6.2]; Exercise 5 is in Bouldin [11, Theorem 1]; Exercise 6 is in Bhatia [8, Theorem 1]; and Exercise 7 is in Khalil and Maher [25, Theorem 3.1] as is Exercise 8 [25, Theorem 4.7].

Chapter 6
Unitary, Isometric and Partially Isometric Approximation of Positive Operators

This chapter is about approximation of positive operators by operators that in some sense preserve size: by—in ascending order of generality—unitaries, isometries and partial isometries.

Rather than take the theory of spectral approximation, given in Chap. 5, as the starting point of this chapter, we build this chapter from the simplest possible finite-dimensional case. This chapter has a strongly geometrical character which, perhaps, gives insight into the isometric, and partially isometric, approximation problems. Further, starting with the finite-dimensional case shows how the problem of finding a unitary approximant arises from, and its solution justifies, a theoretical procedure called the Lowdin orthogonalization, important in quantum chemistry.

In Sects. 6.2 and 6.3 we search for global minimizers amongst the local extrema (cf. the strategy outlined in Chap. 1); and in Sects. 6.2 and 6.3 we use the theory of spectral approximants, Theorem 5.2.6(b), to guarantee the existence of partially isometric global minimizers.

6.1 Quantum Chemical Background: The Lowdin Orthogonalization

Suppose that the quantum chemical system under consideration is decomposed into subsystems (cf. Chap. 3). Suppose that information about these known subsystems is represented by a basis $\{f_1, \ldots, f_n\}$ of unit vectors of $H = \mathbb{C}^n$ (in quantum mechanics all states are represented by unit vectors cf. Sect. 4.1). In order to perform computational and theoretical studies it is necessary to replace the basis $\{f_1, \ldots, f_n\}$ by an **orthogonal** basis of unit vectors. This is done by a matrix B, that transforms the old basis to the new one:

$$Bf_i = e_i, \quad \langle e_i, e_j \rangle = \delta_{ij}, \quad \|f_i\| = 1, \quad i, j = 1, \ldots, n. \tag{6.1.1}$$

© Springer International Publishing AG 2017
P.J. Maher, *Operator Approximant Problems Arising from Quantum Theory*,
DOI 10.1007/978-3-319-61170-9_6

(The term orthogonalization is used, in this context, interchangeably: to refer both to the matrix B and to the basis $\{e_i\}$.)

Since changing the $\{f_i\}$ results in a loss of information it is advisable to minimize the "distance" between the old $\{f_i\}$ and the new $\{e_i\}$ bases. Interpreting distance in the obvious least squares sense, we wish to minimize

$$\sum_{i=1}^{n} \|f_i - e_i\|^2$$

where $Bf_i = e_i$ and the bases $\{e_i\}$ and $\{f_i\}$ are as in (6.1.1).

We reformulate this problem in matricial terms. Since B is invertible it has the unique polar decomposition $B = UL$ where U is unitary and $L = |B|$ is strictly positive (that is, L is invertible and positive). Now,

$$\left.\begin{aligned}
\sum_{i=1}^{n} \|f_i - e_i\|^2 &= \sum_{i=1}^{n} \|B^{-1}e_i - e_i\|^2 \\
&= \sum_{i=1}^{n} \|(L^{-1} - U)U^{-1}e_i\|^2 \\
&= \|L^{-1} - U\|_2^2
\end{aligned}\right\} \qquad (6.1.2)$$

the last equality following from (2.4.4) since $\{U^{-1}e_i\}$ is an orthonormal basis.

Replace B by another orthogonalization $\tilde{B}$, say, where $\tilde{B}f_i = \tilde{e}_i$ and $\langle \tilde{e}_i, \tilde{e}_j \rangle = \delta_{ij}$. Define the operator V by $Ve_i = \tilde{e}_i$. Then V is unitary. Since

$$f_i \xmapsto{UL} e_i \xmapsto{V} \tilde{e}_i$$

the invertible operator $\tilde{B} = V(UL) = (VU)L$; that is, $\tilde{B} = (VU)L$ is the unique polar decomposition of $\tilde{B}$ (with VU unitary). Therefore, $L = |B| = |\tilde{B}|$ is independent of the particular orthogonalization B: varying $B = UL$ is equivalent to varying U over the group of unitary operators in $\mathcal{L}(\mathbb{C}^n)$. Hence, in view of (6.1.2),

$$\min_{B} \sum_{i=1}^{n} \|B^{-1}e_i - e_i\|^2 = \min_{U\text{ unit}} \|L^{-1} - U\|_2^2.$$

Write $A = L^{-1}$ (so that A is strictly positive); then we wish to approximate A by a unitary operator. Theorem 6.1.1 solves this problem by elementary means. More significantly, Theorem 6.1.1 contains a local result which foreshadows some of the later local results (e.g. Theorems 6.2.4, and 6.3.4) of this chapter.

Theorem 6.1.1 *Let A be a strictly positive operator in $\mathcal{L}(\mathbb{C}^n)$. Then:*

(a) for all unitary operators U in $\mathcal{L}(\mathbb{C}^n)$

$$\|A - I\|_2 \le \|A - U\|_2$$

with equality if and only if $U = I$;
(b) V is a local minimizer of the map

$$F_2 : U \mapsto \|A - U\|_2^2,$$

for unitary U, if and only if $V = I$.

Proof

(a) Evaluate $\|A - U\|_2^2 = \sum_{i=1}^{n} \|(A - U)\phi_i\|^2$ over an orthonormal basis $\{\phi_i\}$ consisting of eigenvalues of A with $A\phi_i = \alpha_i \phi_i$ where $\alpha_i > 0$ for $i = 1, \ldots, n$ (by (2.4.4) the $\| \cdot \|_2$ norm is independent of the particular basis $\{\phi_i\}$ chosen). Then, as U is unitary,

$$
\left.
\begin{aligned}
\|A - U\|_2^2 &= \sum_{i=1}^{n} \alpha_i^2 - 2\alpha_i \mathcal{R}\,\langle U\phi_i, \phi_i \rangle + 1 \\
&\ge \sum_{i=1}^{n} \alpha_i^2 - 2\alpha_i + 1 \\
&= \sum_{i=1}^{n} (\alpha_i - 1)^2 \\
&= \|A - I\|_2^2
\end{aligned}
\right\}
\tag{6.1.3}
$$

where the inequality above follows from $\mathcal{R}\,\langle U\phi_i, \phi_i \rangle \le 1$, the latter inequality following from Cauchy–Schwarz.

If $\|A - I\|_2 = \|A - U\|_2$ then equality holds throughout (6.1.3) and hence $\mathcal{R}\,\langle U\phi_i, \phi_i \rangle = 1 = \langle \phi_i, \phi_i \rangle$ for $i = 1, \ldots, n$, that is, since U is unitary, $\langle U\phi_i, \phi_i \rangle = \langle \phi_i, \phi_i \rangle$ for $i = 1, \ldots, n$ so that $U = I$.

(b) requires a heavier hand. Consider the derivative of $F_2 : U \mapsto \|A - U\|_2^2$ at $U = V$. Let H be self-adjoint so that e^{itH}, where $t \in \mathbb{R}$, is unitary. Let $B(t)$ and $f_2(t)$ be defined as follows:

$$B(t) = |A - e^{itH}V|^2,$$

$$f_2(t) = F_2(e^{itH}V) = \tau[B(t)]$$

(cf. (2.4.3)). To evaluate F_2', and hence f_2', we evaluate B'. It can be checked that

$$B(t + s) - B(t) = V^* e^{-itH}(I - e^{-isH})A + Ae^{itH}(I - e^{isH})V$$

so that, on letting $s \to 0$,

$$B'(t) = 2\mathcal{I}m(Ae^{itH}HV).$$

Similarly,

$$B''(t) = 2\mathcal{R}(Ae^{itH}H^2V).$$

By the linearity of trace

$$f_2'(t) = \tau[B'(t)] = 2\tau[\mathcal{I}m(Ae^{itH}HV)],$$
$$f_2''(t) = \tau[B''(t)] = 2\tau[\mathcal{R}(Ae^{itH}H^2V)].$$

Let F_2 have a local minimizer at $U = V$. Then $f_2'(0) = 0$ and $f_2''(0) > 0$. Hence,

$$\tau[\mathcal{I}m(AHV)] = 0 \qquad\qquad (6.1.4)$$

$$\tau[\mathcal{R}(AH^2V)] > 0. \qquad\qquad (6.1.5)$$

To evaluate the traces above, take an orthonormal basis $\{e_1, \ldots, e_n\}$ of eigenvectors of V and let $Ve_i = v_i e_i$ where $|v_i| = 1$ for $i = 1, \ldots, n$. As H can be any self-adjoint operator, take H as the rank 1 projection onto $\mathcal{S}(e_j)$ for some fixed j, that is, $He_i = \delta_{ji} e_i$ for $i = 1, \ldots, n$. Then $AHVe_i = \delta_{ji} v_j Ae_j$ for $i = 1, \ldots, n$. Thus, (6.1.4) says that

$$0 = \frac{1}{2i} \sum_{i=1}^{n} \langle (AHV - V^*HA)e_i, e_j \rangle$$

$$= \frac{1}{2i} (v_j \langle Ae_j, e_j \rangle - \bar{v}_j \langle Ae_j, e_j \rangle)$$

$$= (\mathcal{I}mv_j) \langle Ae_j, e_j \rangle.$$

Since A is strictly positive, $\langle Ae_j, e_j \rangle > 0$ for each $j = 1, \ldots, n$ so that $\mathcal{I}mv_j = 0$, that is, $v_j \in \mathbb{R}$ so that $v_j = \pm 1$ (since $|v_j| = 1$). As V is a local minimizer then from (6.1.5) (since $H^2 = H$ for the projection H) we get, similarly, since $v_j \in \mathbb{R}$,

$$0 < \tau[\mathcal{R}(AHV)] = v_j \mathcal{R} \langle Ae_j, e_j \rangle.$$

Thus, because A is strictly positive, $v_j = 1$. Since this holds for each j then $V = I$.

Conversely, since $V = I$ is a global minimizer, by (a) it is a local minimizer. $\square$

Observe that the proof of the local result, Theorem 6.1.1(b), eschews the use of Theorem 2.4.4 (giving the derivative of $X \mapsto \|X\|_p^p$); instead, it uses ordinary, single-variable calculus.

One can similarly prove the following (finite-dimensional) result about maximization, rather than minimization: for strictly positive A and for all unitaries U in $\mathcal{L}(\mathbb{C}^n)$

$$\|A - U\|_2 \leq \|A + I\|_2 \tag{6.1.6}$$

with equality if and only if $U = -I$; and that V is a local maximizer of the map $U \to \|A - U\|_2^2$ if and only $V = -I$.

6.2 Isometric Approximation of Positive Operators

In this section we minimize

$$\|A - U\|_p,$$

for fixed positive A and varying isometric U such that $A - U \in \mathcal{C}_p$. Most of the work described here was originally obtained for unitary U; but, as we shall see (Lemma 6.2.3 below), there is no distinction in this context between isometric and unitary U: for positive A and U such that $A - U \in \mathcal{C}_p$, where $0 < p \leq \infty$, then U is isometric if and **only if** U is unitary (recall: $\mathcal{C}_\infty$ denotes the two-sided ideal of compact operators; $\|\cdot\|_\infty$ when used, denotes the supremum norm $\|\cdot\|$ on $\mathcal{L}(H)$). For $\|\cdot\|$ on $\mathcal{L}(H)$, we have the following result on (non-normal) isometric approximation (cf. Theorem 6.1.1(a) and (6.1.6)). Observe that no compactness assumptions are made.

Theorem 6.2.1 *Let A be a positive operator. Then for all isometric operators in $\mathcal{L}(H)$*

$$\|A - I\| \leq \|A - U\| \leq \|A + I\|.$$

Proof On the one hand, for isometric U, if $\|f\| = 1$,

$$\|(A - U)f\|^2 = \|Af\|^2 + 1 - 2\mathcal{R} \langle Af, Uf \rangle \geq (\|Af\| - 1)^2$$

since $\mathcal{R} \langle Af, Uf \rangle \leq \|Af\|$ by Cauchy–Schwarz (cf. proof of Theorem 6.1.1(a), inequality (6.1.3)). As A is positive,

$$\inf_{\|f\|=1} \|Af\| = \inf_{\|f\|=1} \langle Af, f \rangle, \quad \sup_{\|f\|=1} \|Af\| = \sup_{\|f\|=1} \langle Af, f \rangle;$$

therefore,

$$\|A - U\| \geq \sup_{\|f\|=1} |\|Af\| - 1|$$

$$= \sup_{\|f\|=1} |\langle Af, f \rangle - 1|$$

$$= \sup_{\|f\|=1} |\langle (A - I)f, f \rangle|$$

$$= \|A - I\|.$$

On the other hand,

$$\|A - U\| = \sup_{\|f\|=1} \|Af - Uf\|$$

$$\leq \sup_{\|f\|=1} (\|Af\| + 1)$$

$$= \sup_{\|f\|=1} \langle (A + I)f, f \rangle$$

$$= \|A + I\|.$$

□

We shall prove that, for $1 < p < \infty$, an isometric approximant, in $\| \cdot \|_p$, of a positive operator A is, likewise, I. We first prove that if there exists an isometric operator U such that $A - U \in C_p$, for $0 < p \leq \infty$, then I is, at least, a candidate for such an approximant of A, that is $A - I \in C_p$.

Lemma 6.2.2 *Let C be a 2-sided ideal of $\mathcal{L}(H)$. If $A - U \in C$ for some positive operator A and isometric operator U, then $A - I \in C$; in particular, if $A - U \in C_p$, where $0 < p \leq \infty$, then $A - I \in C_p$.*

Proof Now, since $U^*U = I$ then

$$A^2 - I = (A - U^*)(A + U) - AU + U^*A$$

$$= (A - U^*)(A + U) - (A - U^*)U + U^*(A - U). \tag{6.2.1}$$

Since every 2-sided ideal of $\mathcal{L}(H)$ is self-adjoint, if $A - U \in C$ then $A - U^* \in C$ and hence, by (6.2.1), $A^2 - I \in C$. Since A is positive, $A + I$ is invertible and so

$$A - I = (A^2 - I)(A + I)^{-1} \in C.$$

□

In the case when $C = C_\infty$ (the compacts) the above result has a quicker proof which depends on (the Calkin algebra) $\mathcal{L}(H)/C_\infty$ being a C^* algebra: for then, as

$A - U \in \mathcal{C}_\infty$, the image of U under the canonical homomorphism $\pi : \mathcal{L}(H) \to \mathcal{L}(H)/\mathcal{C}_\infty$ is positive and unitary and so is the identity in $\mathcal{L}(H)/\mathcal{C}_\infty$.

In finite dimensions an operator is isometric if and only if it is unitary. The next result (Lemma 6.2.3) says that for positive A such that $A - U$ is compact then U is isometric if **and only if** it is unitary. The proof of Lemma 6.2.3 uses the " Fredholm alternative" (Sect. 2.3, third paragraph) : this says that for compact K a non-zero complex number λ is either an eigenvalue of K or $K - \lambda I$ is invertible.

Lemma 6.2.3 *Let A be a positive operator and U be an isometric operator such that $A - U \in \mathcal{C}_p$, where $0 < p \le \infty$. Then U is unitary.*

Proof From Lemma 6.2.2 it follows that $A - I \in \mathcal{C}_p$. So, $U - I = (A - I) - (A - U) \in \mathcal{C}_p$. Hence, $U - I = K$, say, is compact. Now, $U = I + K$ is isometric and hence one-to-one. So, $\mathrm{Ker}(I + K) = \{0\}$ and hence $-1 (= "\lambda")$ is not an eigenvalue of the compact operator K (for otherwise, $\mathrm{Ker}(I + K)$ would contain a non-zero eigenvector with corresponding eigenvalue -1). Therefore, by the Fredholm alternative, $K - (-1)I (= U)$ is invertible and hence unitary.   □

The relevant global inequality—that I is an isometric approximant in $\| \cdot \|_p$ of a positive operator—follows directly (in Theorem 6.2.6) from the theory of spectral approximants; but to generalize Theorem 6.1.1 fully we need to

1. show that there is no other approximant;
2. obtain the local analogue of Theorem 6.1.1.

We therefore classify the critical points of $F_p : U \mapsto \|A - U\|_p^p$ for varying isometric U and fixed, positive A (in Theorem 6.2.4). To obtain the uniqueness assertion in Theorem 6.2.6 we search amongst the critical points of $U \mapsto \|A - U\|_p^p$ for the global extrema (cf. the strategy sketched in Chap. 1). Compare the proof below with that of the corresponding result for self-adjoint approximation (Theorem 3.2.4).

Theorem 6.2.4 *Let A be positive and let $\mathcal{U}$ be defined by*

$$\mathcal{U} = \{U : U \text{ is isometric and } A - U \in \mathcal{C}_p \text{ where } 1 < p < \infty\}.$$

If $\mathcal{U} \ne \emptyset$, let $F_p : \mathcal{U} \to \mathbb{R}^+$ be defined by

$$F_p : U \mapsto \|A - U\|_p^p.$$

If V is a critical point of F_p then:

(a) $V^*A = AV$ and $VA = AV^*$;
(b) $\mathrm{Ker}A$ reduces V;
(c) $AV = VA$;
(d) the restriction of V to $(\mathrm{Ker}A)^\perp$ is self-adjoint;
(e) V is self-adjoint if A is strictly positive.

Proof

(a) The proof of (a) is the longest: results (b)–(e) are simple deductions from it.

Let V be a critical point of the map F_p. Then, by Lemma 6.2.3, the operator V, being isometric, is unitary. For an arbitrary unit vector z and an arbitrary real θ let the operator $W_z(\theta)$ be defined by

$$W_z(\theta) = e^{i\theta}(z \otimes z) + I - (z \otimes z).$$

(Thus, $W_z(\theta)$ multiplies the z component of every vector by $e^{i\theta}$ and acts like the identity on its orthogonal complement.) So, $W_z(\theta)$ is unitary and hence $VW_z(\theta)$ is unitary. As V is a critical point of F_p then, for each z,

$$\frac{d}{d\theta}(F_p(VW_z(\theta)))\bigg|_{\theta=0} = 0$$

(cf. the proof of Theorem 6.1.1(b) where the critical point V is multiplied by an exponential). Applying the chain rule to the maps

$$\theta \mapsto VW_z(\theta) \mapsto F_p(VW_z(\theta))$$

and using Theorem 2.4.4 we get

$$0 = \frac{d}{d\theta}(F_p(VW_z(\theta)))\bigg|_{\theta=0} = p\mathcal{R}\tau[|A - V|^{p-1}U_1^*Vi(z \otimes z)] \qquad (6.2.2)$$

where $A - V = U_1|A - V|$ is the polar decomposition of $A - V$ (so that $\mathrm{Ker}U_1 = \mathrm{Ker}|A - V|$). From (6.2.2), by (2.4.5)

$$0 = \mathcal{R}i\left\langle |A - V|^{p-1}U_1^*Vz, z\right\rangle$$

that is $\left\langle |A - V|^{p-1}U_1^*Vz, z\right\rangle$ is real. Since this holds for arbitrary z it follows that $|A - V|^{p-1}U_1^*V$ is self-adjoint:

$$V^*U_1|A - V|^{p-1} = |A - V|^{p-1}U_1^*V.$$

As U_1 is a partial isometry and V^* is unitary then V^*U_1 is a partial isometry and

$$\mathrm{Ker}(V^*U_1) = \mathrm{Ker}U_1 = \mathrm{Ker}|A - V| = \mathrm{Ker}|A - V|^{p-1}.$$

Thus, $V^*U_1|A - V|^{p-1}$ exhibits the unique polar decomposition of a self-adjoint operator. Therefore, by Theorem 2.2.1(e), V^*U_1 is self-adjoint (so that

$V^* U_1 = U_1^* V$) and commutes with every power of $|A-V|^{p-1}$, in particular with $|A - V|$. So:

$$V^*(A - V) = V^* U_1 |A - V|$$
$$= |A - V| V^* U_1$$
$$= |A - V| U_1^* V$$
$$= (A - V)^* V.$$

Therefore, $V^* A = AV$.

Since the operator A is self-adjoint, as is the norm $\| \cdot \|_p$, it follows that V is a critical point of $F_p : U \to \|A - U\|_p^p$ if and only if V^* is. Hence, $VA = AV^*$.

(b) $\mathrm{Ker} A$ is invariant under V; for if $f \in \mathrm{Ker} A$ then $Vf \in \mathrm{Ker} A$ because, by (a), $AVf = V^* Af = 0$. Similarly, $\mathrm{Ker} A$ is invariant under V^* (because $AV^* = VA$). Therefore, $\mathrm{Ker} A$ reduces V.

(c) From (a), $A^2 V = AV^* A = VA^2$. Hence, as A is positive, $AV = VA$.

(d) From (a) and (c), $V^* A = AV = VA$. So, $(V^* - V)A = 0$, that is, V is self-adjoint on $\mathrm{Ran} A = (\mathrm{Ker} A)^{\perp}$. As $\mathrm{Ker} A$ reduces V (by (b)) this means that the restriction of V to $(\mathrm{Ker} A)^{\perp}$ is self-adjoint.

(e) This follows from (d) since if A is strictly positive then $(\mathrm{Ker} A)^{\perp} = H$.

□

Corollary 6.2.5 *Under the same hypothesis of Theorem 6.2.4 the same conclusions hold for $0 < p \leq 1$ provided, additionally, that the underlying space H is finite-dimensional and that $A - V$ is invertible.*

Proof The additional hypotheses are required, by Theorem 2.4.4, to ensure the differentiability at V of the map F_p for $0 < p \leq 1$. Otherwise, (since, e.g., $\| \cdot \|_p$ is self-adjoint for $0 < p \leq 1$) the proof is exactly the same as that of Theorem 6.2.4.

□

Recall that a self-adjoint unitary operator V is called a **symmetry** and can be expressed in terms of orthogonal projections E and F thus:

$$V = E - F \text{ where } E + F = I \text{ and } EF = 0 \qquad (6.2.3)$$

[1, p. 62]. Thus, Theorem 6.2.4(e), (c) says that for strictly positive A, the map $F_p : U \mapsto \|A - U\|_p^p$, where $1 < p < \infty$ and U is unitary, has a critical point at a symmetry commuting with A.

Theorem 6.2.6 *Let A be positive and let $\mathcal{U}$ be defined by*

$$\mathcal{U} = \{U : U \text{ is isometric and } A - U \in \mathcal{C}_p \text{ when } 1 \leq p < \infty\}$$

If $U \neq \emptyset$, let $F_p : U \mapsto \mathbb{R}^+$ be defined by $F_p : U \mapsto \|A - U\|_p^p$. Then:

(a) *F_p has a global minimizer at $U = I$ that is unique provided A is strictly positive and $1 < p < \infty$:*

$$\|A - I\|_p \leq \|A - U\|_p, \quad 1 \leq p < \infty \tag{6.2.4}$$

with equality occurring in (6.2.4) if, and for strictly positive A and $1 < p < \infty$ only if, $U = I$;

(b) *if the underlying space H is finite-dimensional, F_p has a global maximizer at $U = -I$ that is unique provided A is strictly positive and $1 < p < \infty$:*

$$\|A - U\|_p \leq \|A + I\|_p, \quad 1 \leq p < \infty \tag{6.2.5}$$

with equality occurring in (6.2.5) if, and for strictly positive A and $1 < p < \infty$ only if, $U = -I$.

(c) *for strictly positive A and $1 < p < \infty$, F_p has a unique local minimizer at $U = I$; and, if H is finite-dimensional, also a unique local maximizer at $U = -I$.*

Proof For $p = \infty$ the inequalities (6.2.4) and (6.2.5) have already been proved in Theorem 6.2.1 for operators A, U and I in $\mathcal{L}(H)$.

(a) Let $1 \leq p < \infty$ and U be varying isometry such that $A - U \in \mathcal{C}_p$. Then, by Lemma 6.2.3, U is unitary. Let $E = \{z : |z| = 1\}$. If a normal operator has its spectrum in E then it is unitary and conversely. Let F be given by

$$F(z) = \begin{cases} \dfrac{z}{|z|} & \text{if } z \neq 0 \\ 1 & \text{if } z = 0 \end{cases}$$

so that F is retraction onto E. Then, for positive A, $F(A) = I$. The inequality (6.2.4) now follows from Theorem 5.2.6(b) for the balanced set E.

To prove uniqueness, let $1 < p < \infty$. Let V be a critical point of F_p (possible if V is a global minimizer). Hence, A and V satisfy Theorem 6.2.4(c) and, as A is self-adjoint and V normal, it follows that $A - V$ and $A - I$ are commuting, compact normal operators. Therefore they have a common orthonormal basis $\{\phi_i\}$ of eigenvectors. So, $\{\phi_i\}$ is a basis of eigenvectors of $V - I(= (A - I) - (A - V))$ and hence of A. Let $\alpha_i = \langle A\phi_i, \phi_i \rangle$ and $\mu_i = \langle V\phi_i, \phi_i \rangle$ so that $\alpha_i \geq 0$ and $|\mu_i| = 1$. Then, by (2.4.6),

$$\left. \begin{aligned} \|A - V\|_p^p &= \sum_{i=1}^{\infty} |\alpha_i - \mu_i|^p \\ &\geq \sum |\alpha_i - 1|^p \\ &= \|A - I\|_p^p \end{aligned} \right\} \tag{6.2.6}$$

the inequality in (6.2.6) following as in the proof of (6.1.3) of Theorem 6.1.1.

Suppose, now A is strictly positive and V is a global minimizer of F_p. Then equality holds throughout (6.2.6); and, as $\alpha_i > 0$, this forces $\mu_i = 1$ for all i, that is, $V = I$.

(b) Suppose first that $1 < p < \infty$. Let H finite-dimensional, of dimension n, then the set of unitaries (isometries) forms a compact group and therefore F_p attains its extreme values. Let W be a global maxima, and hence a critical point, of F_p. Then, as in (a), there exists a common orthonormal basis of eigenvectors $\{\psi_i\}$ of A and W, with corresponding eigenvalues α_i and w_i where $\alpha_i \geq 0$ and $|w_i| = 1, 1 \leq i \leq n$. Then for all isometries U

$$
\left.
\begin{aligned}
\|A - U\|_p^p &\leq \|A - W\|_p^p \\
&= \sum_{i=1}^{n} |\alpha_i - w_i|^p \\
&\leq \sum_{i=1}^{n} |\alpha_i + 1|^p \\
&= \|A + I\|_p^p.
\end{aligned}
\right\}
\tag{6.2.7}
$$

For uniqueness, let A be strictly positive and suppose that W is a global maximizer; it then follows from (6.2.7), as in (a), that $w_i = -1$ for each i, that is, $W = -I$.

For finite dimensions, the extension of the inequality (6.2.6) from $1 < p < \infty$ to $1 \leq p < \infty$ follows from the fact that, for fixed $X, \|X\|_p$ is a right-continuous function of p.

(c) To prove there are no other critical points of F_p we require, in view of Theorem 6.2.4(c), (e), to check every symmetry commuting with A. Let $V(\neq \pm I)$ be a symmetry commuting with A. Then, by (6.2.3), $V = E - F$ where E and F are projections commuting with A and such that $E + F = I$ and $EF = 0$. Now take $U(\neq V)$ as any unitary operator which acts like the identity on the range of E and commutes with E (so that $EUE = UE = E$). Then, as $EF = 0$, hence $FAE = FUE = 0$, we find, on using the identity

$$
\|X\|_p^p = \|EXE\|_p^p + \|EXF\|_p^p + \|FXE\|_p^p + \|FXF\|_p^p,
$$

that

$$
F_p(U) - F_p(V) = \|FAF - FUF\|_p^p - \|FAF + F\|_p^p
$$

(where $V = E - F$). Thus, the global result, applied to the range of F, shows that $F_p(U) - F_p(V) < 0$ so that V is not a local maximum. Similarly, V is not a local minimum.

$\square$

The hypothesis in (b) and (c) above that H is finite-dimensional is necessary since in infinite dimensions the conditions that $A + I \in C_p$ and $A \geq 0$ are incompatible.

The following example shows that in the $p = 1$ case there may be more than one minimizer, even for strictly positive A.

Example 6.2.7 Let $H = \mathbb{C}^2$ and A and U be given by

$$A = \begin{bmatrix} 1 + \alpha & 0 \\ 0 & 1 - \alpha \end{bmatrix}, \text{ where } 0 < \alpha < 1, \text{ and } U = \begin{bmatrix} \cos\theta & -\sin\theta \\ \sin\theta & \cos\theta \end{bmatrix}.$$

It can be shown that $\|A - U\|_1 = 2\alpha$. As $\|A - I\|_1 = 2\alpha$ therefore $\|A - I\|_1 = \|A - U\|_1$ without $U = I$.

We state a result that applies to the case $0 < p < 1$.

Theorem 6.2.8 *Let A be strictly positive and $\mathcal{U}$ be defined by*

$$\mathcal{U} = \{U : U \text{ is isometric and } A - U \in C_p \text{ where } 0 < p < \infty\}.$$

If $\mathcal{U} \neq \emptyset$, let $F_p : \mathcal{U} \to \mathbb{R}^+$ be defined by $F_p : U \mapsto \|A - U\|_p^p$. Then:

(a) $F_p(U) \geq F_p(I)$ for all U in $\mathcal{U}$ such that $AU = UA$;
(b) if $A > I$ or if $A < I$ (that is, if the unit circle does not separate the spectrum of A) then, for all U in $\mathcal{U}$, $F_p(U) \geq F_p(I)$ and, if H is finite-dimensional, $F_p(U) \leq F_p(-I)$.

Proof See Exercise 5.   □

We finally comment that the positivity condition on A can be weakened. Let A be an operator such that $\dim\mathrm{Ker}A = \dim\mathrm{Ker}A^*$ (which occurs if, for instance, A is normal or automatically in finite dimensions): then $A = \hat{U}_0|A|$ where $\hat{U}_0$ is unitary (by Theorem 2.2.1(d)). If U is unitary and such that $A - U \in C_p$ for $1 \leq p < \infty$; then $\hat{U}_0^* U$ is unitary and the equality

$$\|A - U\|_p = \||A| - \hat{U}_0^* U\|_p \tag{6.2.8}$$

shows (on applying Theorem 6.2.6 to the R.H.S of (6.2.8)) that the conclusion of Theorem 6.2.6 holds with $\hat{U}_0$ replacing I (thus, for $1 \leq p < \infty$, $\hat{U}_0$ is a unitary approximant in $\|\cdot\|_p$ of A). The same result holds if U is assumed isometric: for then $\hat{U}_0^* U$ is isometric and, as $|A|$ is positive, Lemma 6.2.3 applies: $\hat{U}_0^* U$, and hence U, is unitary (so that $\hat{U}_0$ is an isometric (because unitary) approximant in $\|\cdot\|_p$ of A). Obviously, there is a corresponding result in the supremum norm: if A satisfies $\dim\mathrm{Ker}A = \dim\mathrm{Ker}A^*$ and if $\hat{U}_0$ is as above then

$$\|A - \hat{U}_0\| \leq \|A - U\| \leq \|A + \hat{U}_0\|$$

for all isometries U in $\mathcal{L}(H)$.

Finally, since the norms $\| \cdot \|_p$ and $\| \cdot \|$ are self-adjoint all the results adduced so far on isometric approximation hold for co-isometric approximation (U is a co-isometry if U^* is an isometry).

6.3 Partially Isometric Approximation of Positive Operators

Approximating a positive operator by a partial isometry is more complicated (even in finite dimensions) than approximating by an isometry. This is inevitable: for a partial isometry U comes equipped with an initial space $(\mathrm{Ker}U)^\perp$. Hence, when varying U we must consider the variation of its initial space (whereas no such issue arises in isometric approximation: the "initial space" of an isometry is all of H).

It is therefore hardly surprising that a partially isometric approximant of some positive operator is not necessarily (the partial isometry) I. This is illustrated below.

Example 6.3.1 Let $H = \mathbb{C}^2$ and A, U and I be given by

$$A = \begin{bmatrix} \sqrt{2} & \sqrt{2} \\ \sqrt{2} & 2 \end{bmatrix}, \quad U = \begin{bmatrix} \frac{1}{2} & \frac{1}{2} \\ \frac{1}{2} & \frac{1}{2} \end{bmatrix}, \quad I = \begin{bmatrix} 1 & 0 \\ 0 & 1 \end{bmatrix}.$$

Then A is positive, U (the projection onto the line $y = x$) is partially isometric and

$$\|A - U\|_2 < \|A - I\|_2.$$

Nevertheless, for a given, specific partial isometry U, the operator that acts like the identity on its initial space, viz. $E_U = U^*U$ (the projection onto its initial space $(\mathrm{Ker}U)^\perp$) shares some similarities with the identity in isometric approximation. Equally, $F_U = UU^*$ (the projection onto the final space, $\mathrm{Ran}U$) plays a similar role to I in isometric approximation since it acts like the identity on $\mathrm{Ran}U$.

Thus, Lemma 6.3.2 is analogous to Lemma 6.2.2. Notice that Lemma 6.3.2 is stated for self-adjoint, rather than positive A: the positivity of A is only used in the proof of Lemma 6.2.2 to ensure the invertibility of $A + I$.

Lemma 6.3.2 *Let $\mathcal{C}$ be a 2-sided ideal of $\mathcal{L}(H)$. If $A - U \in \mathcal{C}$ for some self-adjoint operator A and some partial isometry U then $A^2 - E_U \in \mathcal{C}$ and $A^2 - F_U \in \mathcal{C}$ (where $E_U = U^*U$ and $F_U = UU^*$); in particular, if $A - U \in \mathcal{C}_p$, where $0 < p < \infty$, then $A^2 - E_U \in \mathcal{C}_p$ and $A^2 - F_U \in \mathcal{C}_p$*

Proof It follows (as in the proof of Lemma 6.2.2) from the self-adjointness of A and of $\mathcal{C}$ that

$$A^2 - E_U = (A - U^*)(A + U) - AU + U^*A \in \mathcal{C}.$$

Similarly, $A^2 - F_U \in \mathcal{C}$. $\square$

Lemma 6.3.3 below says further that if, for positive A and partially isometric U, the operator $A - U$ is compact then so, too, is $A - E_U$ (and, similar, $A - F_U$).

Lemma 6.3.3 *Let A be positive and U be a partial isometry such that $A - U \in C_p$ for $0 < p \le \infty$. Then*

$$A^2 - A \in C_\infty, \; A - E_U \in C_\infty \text{ and } A - F_U \in C_\infty.$$

Proof If $A - U \in C_p$, where $0 < p \le \infty$, then by Lemma 6.3.2, $A^2 - E_U \in C_p$ (and $A^2 - F_U \in C_p$). Hence, in particular, $A^2 - E_U \in C_\infty$. If π is the canonical homomorphism on $\mathcal{L}(H)$ onto the Calkin algebra $\mathcal{L}(H)/C_\infty$ then $\pi(A^2) = \pi(E_U)$. Using the homomorphism property of π (specially, that $(\pi(X))^2 = \pi(X^2)$) we then have

$$(\pi(A))^4 = (\pi(E_U))^2 = \pi(E_U) = (\pi(A))^2.$$

Write $a = \pi(A)$. Then $a^4 = a^2$. Taking positive square roots of this (here we use the positivity of A) we get $a^2 = a$, that is, $A^2 - A \in C_\infty$. Hence, $A - E_U = (A - A^2) + (A^2 - E_U) \in C_\infty$. Similarly, $A - F_U \in C_\infty$. $\square$

To find a global minimizer of the map $F_p : U \mapsto \|A - U\|_p^p$, where A is positive and U is partially isometric, we classify as in the isometric case the critical points of the map F_p.

Theorem 6.3.4 *Let A be positive and let $\mathcal{U}$ be defined by*

$$\mathcal{U} = \{U : U \text{ is partially isometric and } A - U \in C_p \text{ where } 1 < p < \infty\}.$$

If $\mathcal{U} \neq \emptyset$, let $F_p : \mathcal{U} \to \mathbb{R}^+$ be defined by

$$F_p : U \mapsto \|A - U\|_p^p.$$

If V is a critical point of F_p then:

(a) $V^*A = AV$ and $VA = AV^*$;
(b) $E_VA = AE_V$;
(c) $\text{Ker}V$ reduces A and $\text{ker}A$ reduces E_V;
(d) $\text{Ker}A$ reduces V;
(e) $AV = VA$;
(f) the restriction of V to $(\text{Ker}A)^\perp$ is self-adjoint;
(g) V is self-adjoint if A is strictly positive.

Proof

(a) The proof is analogous to that of Theorem 6.2.4 for isometric approximation. Thus, for an arbitrary unit vector z and an arbitrary real θ, let the unitary operator $W_z(\theta)$ be defined by

$$W_z(\theta) = e^{i\theta}(z \otimes z) + I - (z \otimes z).$$

Let V be a critical point of F_p. Then, as $W_z(\theta)$ is unitary, $W_z(\theta)V$ and $VW_z(\theta)$ are partial isometries and, for each z,

$$\left.\frac{dF_p(W_z(\theta)V)}{d\theta}\right|_{\theta=0} = 0 = \left.\frac{dF_p(VW_z(\theta))}{d\theta}\right|_{\theta=0}. \qquad (6.3.1)$$

Consider (this time) the operator $W_z(\theta)V$. Applying the chain rule to the maps $\theta \mapsto W_z(\theta)V \mapsto F_p(W_z(\theta)V)$ and using Theorem 2.4.4 we get

$$0 = \left.\frac{dF_p(W_z(\theta)V)}{d\theta}\right|_{\theta=0} = p\mathcal{R}\tau[|A - V|^{p-1}U_1^* i(z \otimes z)V] \qquad (6.3.2)$$

where $A - V = U_1|A - V|$ is the polar decomposition of $A - V$. From (6.3.2) (since $\tau[S(z \otimes z)T] = \langle TSz, z\rangle$ (2.4.5)) it follows (cf. Theorem 6.2.4) that $V|A - V|^{p-1}U_1^*$ is self-adjoint:

$$V|A - V|^{p-1}U_1^* = U_1|A - V|^{p-1}V^*. \qquad (6.3.3)$$

Similarly, from the second equality in (6.3.1) it follows (exactly as in Theorem 6.2.4) that

$$V^*U_1|A - V|^{p-1} = |A - V|^{p-1}U_1^*V. \qquad (6.3.4)$$

Now, we will have, $V^*A = AV$ if and only if

$$V^*U_1|A - V| = |A - V|U_1^*V \qquad (6.3.5)$$

because (6.3.5) is equivalent to $V^*A - V^*V = AV - V^*V$ since $A - V = U_1|A-V|$ and $A^* = A$.

Proof of (6.3.5). If $p = 2$ it is obvious that (6.3.5) holds; for then (6.3.5) is the same as (6.3.4).

For arbitrary p, where $1 < p < \infty$, we use the functional calculus (a variant of the following argument is in the proof of Theorem 3.2.4). Write

$$Y = U_1^*V \text{ and } Z = |A - V|^{p-1}.$$

Then (6.3.4) says that

$$Y^*Z = ZY \qquad (6.3.6)$$

and (6.3.5) is the same as $Y^*Z^{\frac{1}{p-1}} = Z^{\frac{1}{p-1}}Y$. This will follow by the functional calculus from

$$Y^*Z^n = Z^nY, \quad n \in \mathbb{N}. \qquad (6.3.7)$$

The proof of (6.3.7) is by induction, first for odd n and then for even n. We need the following assertion: $YZ = ZY^*$. The proof of this is as in Theorem 3.2.4: since $\mathrm{Ran}Z = (\mathrm{Ker}U_1)^\perp$ (because $\mathrm{Ker}U_1 = \mathrm{Ker}|A-V| = \mathrm{Ker}Z$ by (2.2.1)) then $U_1^*U_1$, the orthogonal projection onto $(\mathrm{Ker}U_1)^\perp$, satisfies $ZU_1^*U_1 = Z = U_1^*U_1Z$. Multiplying (6.3.3) (viz. $VZU_1^* = U_1ZV^*$) on the left by U_1^* and on the right by U_1 we get

$$U_1^*VZU_1^*U_1 = U_1^*U_1ZV^*U_1,$$

that is, $YZ = ZY^*$ as desired. Returning to (6.3.7) in the case of n odd: for $n = 1$, (6.3.7) is just (6.3.6); whilst the inductive step follows from the assertion (in the form $ZY^* = YZ$) and from (6.3.6). The final step—that (6.3.7) holds for even n, too—follows from another application of the functional calculus. Since (6.3.7) holds for odd n then

$$(Y^*Z^{2k-1})(Z^{2l}) = (Z^{2l})(Z^{2k-1}Y)$$

where $l \geq 0$ and $k \geq 1$. Hence, for every polynomial q, $(Y^*Z^{2k-1})q(Z^2) = q(Z^2)(Z^{2k-1}Y)$. Take a sequence $\{q_i\}$ of polynomials converging uniformly to the positive square root function $t \to t^{1/2}$ where $t \in \mathbb{R}^+$. Then, as $(Y^*Z^{2k-1})q_i(Z^2) = q_i(Z^2)(Z^{2k-1}Y)$ for every q_i, it follows, on taking limits, that $(Y^*Z^{2k-1})Z = Z(Z^{2k-1}Y)$, that is, $Y^*Z^{2k} = Z^{2k}Y$—which is (6.3.7) for even n. This proves $V^*A = AV$.

The relation $VA = AV^*$ follows immediately (as in Theorem 6.2.4(a)) since the operator A and the norm $\|\cdot\|_p$ are self-adjoint.

(b) From (a), $E_VA^2 = A^2E_V$ (for, by (a), $E_VA^2 = V^*(VA)A = V^*(AV^*)A = (AV)(AV) = A(AV^*)V$). Hence, E_V commutes with A (the positive square root of A^2).

(c) $E_VA = AE_V$ means that $\mathrm{Ker}V(= (\mathrm{Ran}E_V)^\perp)$ reduces A. $\mathrm{Ker}A$ reduces E_V since if $f \in \mathrm{Ker}A$ (so that $Af = 0$) then $E_Vf \in \mathrm{Ker}A$ since $E_VA = AE_V$ (by (b)).

(d), (e), (f), (g). The proofs of (d)–(g) are identical to those of Theorem 6.2.4(b)–(e) respectively and so are omitted.   □

Note that the positivity of A, though required in parts (b), (c), (e) and (f) above, is not required in part (a) which holds if A is self-adjoint. Indeed, the differentiation argument (cf. proof of Theorem 6.3.4(a), (6.3.5)), gives the following result: Let A be in $\mathcal{L}(H)$ and let U vary over those partial isometries such that $A - U \in C_p$, where $1 \leq p < \infty$, then if V is a critical point of the map $F_p : U \to \|A - U\|_p$, it follows that $V^*A = A^*V$.

Observe that the proof of the Theorem 6.3.4 (in particular, the argument involving approximating to the function $t \to t^{\frac{1}{p-1}}$, where $t \geq 0$) does not work in the $0 < p \leq 1$ case. Of course, this does not preclude Theorem 6.3.4 from holding when $0 < p \leq 1$ provided F_p is differentiable at V.

The properties adduced in Theorem 6.3.4—and especially the geometric ones—are, as we shall see, crucial in determining a partially isometric approximant of a positive operator.

It is hard to guess what such an approximant might look like. Certainly not the identity, as Example 6.3.1 shows. A plausible guess would be that such an approximant might be some projection—plausible since a projection acts like the identity on some subspace. One way into the problem is to consider the following *very* special case.

A Very Special Case Consider approximating some positive operator A by a projection E that commutes with A. For simplicity take H as being finite-dimensional. Since A and E commute there exists a basis $\{\phi_i\}$ of $H(= \mathbb{C}^n$, say) consisting of eigenvectors of A and E. Let

$$A\phi_i = \alpha_i\phi_i, \quad \text{and } E\,\phi_i = \xi_i\phi_i$$

(so that $\alpha_i \geq 0$ and $\xi_i = 0$ or 1). As an approximant try the projection E_t defined, for some fixed t, as the projection onto the subspace M_t given by,

$$M_t = \mathcal{S}\{\phi_i : \alpha_i \geq t \text{ where } A\phi_i = \alpha_i\phi_i\}$$

(of course, we could, equally well, have taken $\alpha_i > t$—which will explain the uniqueness condition later). The eigenvectors of A and E_t also coincide; and $AE_t = E_tA$. Let

$$E_t\phi_i = e_i\phi_i,$$

We want to choose t so that E_t is closest to A in $\|\cdot\|_2$ (for simplicity). We therefore choose t so that, for each $i = 1, \dots, n$,

$$|\alpha_i - e_i| \leq |\alpha_i - \xi_i|; \tag{6.3.8}$$

since then, by (2.4.6),

$$\|A - E_t\|_2 \leq \|A - E\|_2.$$

If ϕ_i is in RanE and RanE$_t$ or if ϕ_i is in KerE and KerE$_t$ then nothing happens: equality holds in (6.3.8). But if ϕ_i is in KerE and RanE$_t$ then $|\alpha_i - 1| \leq |\alpha_i|$ so that $\frac{1}{2} \leq \alpha_i$ whence $t \geq \frac{1}{2}$; whilst if ϕ_i is in RanE and KerE$_t$ then, similarly, $t \leq \frac{1}{2}$. So, $t = \frac{1}{2}$. Conclusion: if the projections E commute with the positive matrix A then

$$\|A - E_{\frac{1}{2}}\|_2 \leq \|A - E\|_2 \tag{6.3.9}$$

where $E_{\frac{1}{2}}$ is the projection onto the subspace $M_{\frac{1}{2}}$ given by

$$M_{\frac{1}{2}} = S\{\phi_{i_t} : \alpha_i \geq \frac{1}{2} \text{ where } A\phi_i = \alpha_i\phi_i\}$$

(there is also a uniqueness condition, involving $\alpha_i \neq \frac{1}{2}$, proved in Theorem 6.3.6).

The generalization of (6.3.9) to the non-commutative (infinite-dimensional) case of partially isometric approximation is in three main steps. The first step (Theorem 6.3.6) assumes that $U \to \|A - U\|_p^p$, for $1 < p < \infty$ has a critical point at $U = V$ and minimizes $\|A - V\|_p$; the second step (Theorem 6.3.9) assumes that $U \to \|A - U\|_p^p$ has a global minimizer at $U = V$ and minimizes $\|A - V\|_p^p$; the third step (Theorem 6.3.11) proves that $U \to \|A - U\|_p^p$ has a global minimizer.

Theorem 6.3.6 depends crucially on the geometric repercussions of Theorem 6.3.4. Its proof also used the following elementary proposition.

Proposition 6.3.5 *Let K be a compact normal operator in $\mathcal{L}(H)$ and let H be the direct sum $H = \oplus H_n$ of a (possibly countably infinite) number of subspaces H_n, each of which reduces K. Then there exists a basis $\{\phi_i\}$ of H consisting of eigenvectors of K and such that each ϕ_i is in only one H_n.*

Proof Fix n. Since H_n reduces K the restriction of K to H_n, $K|_{H_n}$, is a compact normal operator in $\mathcal{L}(H)$. Hence there exists a basis of H_n consisting of eigenvectors of $K|_{H_n}$. Now let n vary and take the union of all such bases. This union, $\{\phi_i\}$ say, is a basis of H consisting of eigenvectors of K and such that each ϕ_i is in only one H_n.   □

Theorem 6.3.6 *Let A be a positive operator. Let V be a critical point of*

$$F_p : U \to \|A - U\|_p^p$$

where U varies over those partial isometries such that $A - U \in C_p$, where $1 < p < \infty$. Then:

(a) $E_V A = AE_V, A - E_V \in C_p$ and

$$\|A - E_V\|_p \leq \|A - V\|_p \tag{6.3.10}$$

with equality occurring in (6.3.10) if, and for strictly positive A only if, $V = E_V$;
(b) if the underlying space is finite-dimensional

$$\|A - E_V\|_p \leq \|A - V\|_p \leq \|A + E_V\|_p \tag{6.3.11}$$

with equality occurring in the left-hand (right-hand) of (6.3.11) if, and for strictly positive A only if, $V = E_V (V = -E_V)$.

Proof

(a) Let V be a critical point of F_p. Then, by Theorem 6.3.4(b), (c) and (e), $E_V A = A E_V$, $\mathrm{Ker}V$ reduces to A, and hence $A - E_V$, and $AV = VA$. Also, by Lemma 6.3.3, $A - E_V$ is compact since A is positive.

Suppose, firstly, that A is strictly positive (so that $\mathrm{Ker}A = \{0\}$). Then, by Theorem 6.3.4(g), $V = V^*$ so that $A - V$ is reduced by $\mathrm{Ker}V$ and $A - V$ commutes with $A - E_V$. Hence, since the compact normal operators $(A - V)|_{\mathrm{Ker}V}$ and $(A - E_V)|_{\mathrm{Ker}V}$ in $\mathcal{L}(\mathrm{Ker}V)$ commute there exists a basis of $\mathrm{Ker}V$ consisting of common eigenvectors of

$$(A - V)|_{\mathrm{Ker}V} \quad \text{and} \quad (A - E_V)|_{\mathrm{Ker}V} .$$

There is a similar result about common eigenvectors of

$$(A - V)|_{(\mathrm{Ker}V)^\perp} \quad \text{and} \quad (A - E_V)|_{(\mathrm{Ker}V)^\perp} .$$

Applying Proposition 6.3.5 there is, therefore, a basis $\{\phi_i\}$, say, of H consisting of common eigenvectors of $A - V$ and $A - E_V$ and such that either $\phi_i \in \mathrm{Ker}V$ or $\phi_i \in (\mathrm{Ker}V)^\perp$ for each i. Thus, each ϕ_i is an eigenvector of E_V, A and V. Let α_i, ξ_i and v_i be the corresponding eigenvalues of A, E_V and V respectively. Then, for each i,

$$|\alpha_i - v_i| \geq |\alpha_i - \xi_i|; \tag{6.3.12}$$

for if $\phi_i \in \mathrm{Ker}V$ then $v_i = \xi_i = 0$ and if $\phi_i \in (\mathrm{Ker}V)^\perp$ then $\xi_i = 1 = |v_i|$ which, since $\alpha_i \geq 0$, gives the desired inequality (6.3.12). Therefore, as the normal operator $A - V \in \mathcal{C}_p$ then, by (2.4.6),

$$\|A - V\|_p^p = \sum_{i=1}^{\infty} |\alpha_i - v_i|^p \geq \sum_{i=1}^{\infty} |\alpha_i - \xi_i|^p. \tag{6.3.13}$$

Hence, by (2.4.6) again, the normal operator $A - E_V \in \mathcal{C}_p$ and

$$\|A - E_V\|_p^p = \sum_{i=1}^{\infty} |\alpha_i - \xi_i|^p$$

which gives the inequality (6.3.10).

As for the uniqueness assertion, let there be equality in (6.3.10). Then there is equality throughout (6.3.13) and hence, by (6.3.12) $|\alpha_i - v_i| = |\alpha_i - \xi_i|$ for each i. If $v_i = \xi_i = 0$ this equality holds automatically whilst if $|v_i| = 1 = \xi_i$ then $|\alpha_i - v_i| = |\alpha_i - 1|$ which forces $\mathcal{R}v_i = 1$ (because $\alpha_i > 0$ since A is strictly positive) and hence $v_i = 1$. So $v_i = \xi_i$ for each i, that is, $V = E_V$.

Next, we extend the inequality (6.3.10) to positive (as distinct from strictly positive) A. Consider, this time, $\text{Ker}A$. Since, by Theorem 6.3.4(c) and (d), $\text{Ker}A$ reduces E_V and V, therefore $\text{Ker}A$ reduces $A - E_V$ and $A - V$. Decompose $A - V$ into its restrictions to $\text{Ker}A$ and $(\text{Ker}A)^\perp$, viz,

$$(A - V)|_{\text{Ker}A} \ (= S) \text{ and } (A - V)|_{(\text{Ker}A)^\perp} \ (= T, \text{ say})$$

Since $S + T \in \mathcal{C}_p$ and since $\text{Ran}S \perp \text{Ran}T$ and $\text{Ran}S^* \perp \text{Ran}T^*$ it follows that (2.4.8) applies: $S \in \mathcal{C}_p, T \in \mathcal{C}_p$ and

$$\|A - V\|_p^p = \|S\|_p^p + \|T\|_p^p. \tag{6.3.14}$$

Now, $S = (A - V)|_{\text{Ker}A} = -V|_{\text{Ker}A}$ and $|V|^p = |E_V|^p$ and so, since $\|X\|_p^p = \tau[|X|^p]$ if $X \in \mathcal{C}_p$, it follows that

$$\|S\|_p^p = \| \ (A - E_V)|_{\text{Ker}A} \ \|_p^p.$$

As for T, since A is strictly positive on $(\text{Ker}A)^\perp$ the first part of the proof shows that $(A - E_V)|_{(\text{Ker}A)^\perp} \in \mathcal{C}_p$ and that

$$\|T\|_p^p = \| \ (A - V)|_{(\text{Ker}A)^\perp} \ \|_p^p \geq \| \ (A - E_V)|_{(\text{Ker}A)^\perp} \ \|_p^p$$

substituting back into (6.3.14) and again using (2.4.8) we obtain the desired inequality (6.3.10).
(b) The proof is similar to (a) and so is omitted.   □

The next example shows that the inequality $\|A - E_V\|_p \leq \|A - V\|_p$ does not hold for all partial isometries V such that $A - V \in \mathcal{C}_p$.

Example 6.3.7 Let $H = \mathbb{C}^2$ and A (as in Example 6.3.1) and V be given by

$$A = \begin{bmatrix} \sqrt{2} & \sqrt{2} \\ \sqrt{2} & 2 \end{bmatrix}, \quad \text{and} V = \begin{bmatrix} \frac{1}{\sqrt{2}} & 0 \\ \frac{1}{\sqrt{2}} & 0 \end{bmatrix}.$$

Then V is a partial isometry with initial space the x axis and $\|A - V\|_2 \leq \|A - E_V\|_2$.

Theorem 6.3.9 constitutes the second step in generalizing the result (6.3.9): it assumes that $U \to \|A - U\|_p^p$ has a global minimizer at $U = V$ and minimizes $\|A - V\|_p^p$. It incorporates the reasoning of (6.3.9); accordingly we frame the following definition.

Definition 6.3.8 Let A in $\mathcal{L}(H)$ be positive and such that there exists a basis $\{\phi_i\}$ of H consisting of eigenvectors of A. The operator $E_{\frac{1}{2}}$ is defined as the projection onto the subspace $M_{\frac{1}{2}}$ given by

$$M_{\frac{1}{2}} = \mathcal{S}\{\phi_i : \alpha_i \geq \frac{1}{2} \text{ where } A\phi_i = \alpha_i\phi_i\}.$$

Not surprisingly, the same results hold in the rest of this chapter if $E_{\frac{1}{2}}$ is replaced by $E'_{\frac{1}{2}}$ where $E'_{\frac{1}{2}}$ is defined in the same way as $E_{\frac{1}{2}}$ except that the condition $\alpha_i \geq \frac{1}{2}$ is replaced by $\alpha_i > \frac{1}{2}$.

In the statement of Theorem 6.3.9, $\sigma_p(A)$ denotes the point spectrum (that is, the set of eigenvalues) of A.

Theorem 6.3.9 *Let A be a positive operator. Let U vary over those partial isometries such that $A - U \in C_p$ where $1 < p < \infty$. If the map*

$$F_p : U \mapsto \|A - U\|_p^p$$

attains a global minimum then there exists a basis of the underlying space consisting of eigenvectors of A and

$$\|A - E_{\frac{1}{2}}\|_p \leq \|A - U\|_p \tag{6.3.15}$$

where $E_{\frac{1}{2}}$ is as in Definition 6.3.8; and, further, for strictly positive A such that $\frac{1}{2} \notin \sigma_p(A)$, equality occurs in (6.3.15) if and only if $U = E_{\frac{1}{2}}$.

Proof Let F_p attain a global minimum at V, say, so that

$$\|A - V\|_p \leq \|A - U\|_p.$$

Since, for $1 < p < \infty$, a global minimizer is a critical point it follows from Theorem 6.3.6(a) that $E_V A = A E_V, A - E_V \in C_p$ and

$$\|A - E_V\|_p = \|A - V\|_p \leq \|A - U\|_p. \tag{6.3.16}$$

(The equality is because F_p attains a global minimum at V.) The inequality (6.3.15) will now follow, on taking $E = E_V$ from the assertion below.

Assertion Let E be a projection such that $EA = AE$ and $A - E \in C_p$ where $1 < p < \infty$. Then:

(a) there exists a basis $\{\phi_i\}$ of the underlying space consisting of eigenvectors of A and such that $\phi_i \in \text{Ran}E$ or $\phi_i \in (\text{Ran}E)^{\perp}$ for each i;
(b) $A - E_{\frac{1}{2}} \in C_p$ and

$$\|A - E_{\frac{1}{2}}\|_p \leq \|A - E\|_p; \tag{6.3.17}$$

and, provided $\frac{1}{2} \notin \sigma_p(A)$, equality holds in (6.3.17) if and only if $E = E_{\frac{1}{2}}$.

Proof of Assertion

(a) Since $EA = AE$ the compact normal operator $A - E$ is reduced by $\text{Ran}E$. Therefore, as in Proposition 6.3.5, there exists a basis $\{\phi_i\}$ of H consisting of eigenvectors of $A - E$ and such that $\phi_i \in \mathrm{Ran}E$ or $\phi_i \in (\mathrm{Ran}E)^\perp$ for each i. Each such ϕ_i is therefore an eigenvector of E, A and of $E_{\frac{1}{2}}$.

(b) Let $A\phi_i = \alpha_i \phi_i$, $E\phi_i = \xi_i \phi_i$ and $E_{\frac{1}{2}}\phi_i = e_i \phi_i$ for each i. Then, exactly as in the very special case (6.3.8).

$$|\alpha_i - e_i| \leq |\alpha_i - \xi_i|$$

for each i, so that, by (2.4.6),

$$\sum_{i=1}^{\infty} |\alpha_i - e_i|^p \leq \sum_{i=1}^{\infty} |\alpha_i - \xi_i|^p = \|A - E\|_p^p.$$

This proves $A - E_{\frac{1}{2}} \in \mathcal{C}_p$ and gives (6.3.17).

Next, if equality holds in (6.3.17) then $|\alpha_i - e_i| = |\alpha_i - \xi_i|$ for each i. This forces $\mathrm{Ran}E = M_{\frac{1}{2}}$ since $\frac{1}{2} \notin \sigma_p(A)$; for if either $\phi_i \in \mathrm{Ran}E$ and $\phi_i \notin M_{\frac{1}{2}}$ (when $\alpha_i < \frac{1}{2}$) or if $\phi_i \notin \mathrm{Ran}E$ and $\phi_i \in M_{\frac{1}{2}}$ (when $\alpha_i > \frac{1}{2}$) we would have $|\alpha_i - e_i| < |\alpha_i - \xi_i|$. This proves the assertion.

Finally, returning to the theorem, let A be strictly positive and such that $\frac{1}{2} \notin \sigma_p(A)$. If there is equality in (6.3.15) for some partial isometry U then applying (6.3.16)

$$\|A - E_{\frac{1}{2}}\|_p = \|A - E_U\|_p = \|A - U\|_p.$$

The first equality implies, by the assertion, that $E_U = E_{\frac{1}{2}}$; the second equality implies, by Theorem 6.3.6(a), that $U = E_U$. So, $U = E_{\frac{1}{2}}$.   □

The problem of finding a partially isometric approximant of a positive operator thus becomes an existence problem (the third, and final, step of extending (6.3.9)). This is solved, for finite dimensions, in the following theorem.

Theorem 6.3.10 *Let the underlying space H be finite dimensional. Let A be a positive operator and $E_{\frac{1}{2}}$ be as in Definition 6.3.8. Then for all partial isometries U in $\mathcal{L}(H)$*

$$\|A - E_{\frac{1}{2}}\|_p \leq \|A - U\|_p \leq \|A + I\|_p \quad \text{where} \quad 1 \leq p < \infty, \tag{6.3.18}$$

$$\|A - E_{\frac{1}{2}}\|p \leq \|A - U\|p \leq \|A + I\|; \tag{6.3.19}$$

for $1 \leq p < \infty$ and for strictly positive A the right-hand inequality in (6.3.18) is an equality if and only if $U = -I$; and, further, for strictly positive A such that $\frac{1}{2} \notin \sigma_p(A)$ the left hand inequality in (6.3.18) is an equality if and only if $U = E_{\frac{1}{2}}$.

Proof Let $1 < p < \infty$. The set of all partial isometries is closed and bounded [23, Problem 129] and hence, since H is finite-dimensional, compact, it follows, as in [1, Theorem 3.5] that the map $F_p : U \to \|A - U\|$ is bounded and attains its bounds. The left-hand inequality in (6.3.18), and the corresponding uniqueness assertion, now follow from Theorem 6.3.9.

To prove the right-hand inequality in (6.3.18) let W be a global maximum, and hence a critical point, of F_p. Then by Theorem 6.3.6(a), (b) we have $E_W A = A E_W$ and

$$\|A - W\|_p = \|A + E\|_p$$

which, for strictly positive A, forces $W = -E_W$. It can be shown, by considering the eigenvalues of A, that if E is a projection such $EA = AE$ and if H is finite-dimensional then $\|A + E\|_p$ attains its maximum at $E = I$ and at no other point. This gives the right-hand inequality in (6.3.18) and the corresponding uniqueness assertion.

Finally, let $p = 1$ or $p = \infty$. As $\|X\|_p$ is a continuous function of p in finite dimensions, the $p = 1$ inequality (6.3.18) follows from the $1 < p$ inequality (6.3.18)) (on letting $p \to 1$) as does the sup norm inequality (6.3.19) (on letting $p \to \infty$).   □

In finite dimensions the condition on A of positivity can be dropped: in that case $A = \hat{U}_0|A|$ where $\hat{U}_0$ is unitary (cf. Theorem 2.2.1(a)). Let $\{\phi_i\}$ be a basis of H consisting of eigenvectors of $|A|$ and let $\hat{E}_{\frac{1}{2}}$ be the projection onto the subspace

$$\bar{S}\{\phi_i : \alpha_i \geq \frac{1}{2} \text{ when } |A|\phi_i = \alpha_i\phi_i\}.$$

Then if U (and hence $\hat{U}_0^* U$) is a partial isometry it follows, cf. (6.2.8), from Theorem 6.3.10 that $\|A - U\|_p$, where $1 < p < \infty$, is minimized when $U = \hat{U}_0\hat{E}_{\frac{1}{2}}$ and maximized when $U = -\hat{U}_0$. Thus,

$$\|A - \hat{U}_0\hat{E}_{\frac{1}{2}}\|_p \leq \|A - U\|_p \leq \|A + \hat{U}_0\|_p, \text{ where } 1 \leq p < \infty \qquad (6.3.20)$$

(with the now obvious necessary and sufficient conditions for left/right hand equality when $1 < p < \infty$).

We return to the infinite-dimensional case. As for maximizing $\|A - U\|$ it follows, as in Theorem 6.2.1, that if A is positive then for all partial isometries U in $\mathcal{L}(H)$

$$\|A - U\| \leq \|A + I\|. \qquad (6.3.21)$$

To get the infinite-dimensional approximation results we invoke the theory of spectral approximants developed in Chap. 5. First, there is the following result about approximating a normal operator by normal partial isometries.

Theorem 6.3.11 *Let A be normal operator and define the function* $F : E \to \Lambda$, *where* $\Lambda = \{0\} \cup C$ *with* $C = \{z : |z| = 1\}$, *by*

$$F(re^{i\theta}) = \begin{cases} e^{i\theta} & \text{if } r \geq \dfrac{1}{2} \\ 0 & \text{if } r < \dfrac{1}{2}. \end{cases}$$

Then:

(a) $F(A)$ *is a normal partial isometry and for all normal partial isometries* U

$$\|A - F(A)\| \leq \|A - U\|; \qquad (6.3.22)$$

(b) *for all normal partial isometries* U *such that* $A - U \in C_p$, *where* $1 \leq p < \infty$, *it follows that* $A - F(A) \in C_p$ *and*

$$\|A - F(A)\|_p \leq \|A - U\|_p. \qquad (6.3.23)$$

Proof

(a) First, the spectrum of a normal partial isometry is a non-empty closed subset of $\Lambda (= \{0\} \cup C)$. This is because (i) a normal partial isometry is the direct sum of the zero operator and a unitary, and conversely [23, Problem 204]; and (ii) the spectrum of the direct sum of two operators is the union of their individual spectra.

Conversely, if the spectrum of some normal operator, X, say, is a non-empty closed subset of Λ then the underlying space H can be decomposed so that X is the direct sum of a normal quasinilpotent operator e.g. the zero operator, and a unitary. Hence, X is a normal partial isometry (in the notation of Chap. 5, $X \in \mathcal{J}(\Lambda) = \{\text{all normal partial isometries}\}$).

The mapping $F : \mathbb{C} \to \Lambda$ is a retraction onto Λ. Therefore, by Theorem 5.1.4, $\sigma(F(A)) \in \Lambda$, so that $F(A)$ is a normal partial isometry, and $F(A)$ satisfies (6.3.22).

(b) Moreover, Λ is **balanced**. Therefore, Theorem 5.2.6(b) applies : $A - F(A) \in C_p$ and the inequality (6.3.23) holds.   □

Of course, the same results as in Theorem 6.3.11 hold if F is replaced by the function $F' : \mathbb{C} \to \Lambda$ defined in the same way as F except that the condition $r \geq \frac{1}{2}$ $(r < \frac{1}{2})$ is replaced by $r > \frac{1}{2}$ $(r \leq \frac{1}{2})$ (Fig. 6.1).
And now for the final result.

Fig. 6.1 The outer unit circle
and the origin is the spectrum
of a normal partial isometry;
the inner concentric circle of
radius $\frac{1}{2}$ defines the retraction

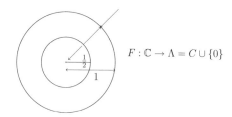

$F : \mathbb{C} \to \Lambda = C \cup \{0\}$

Theorem 6.3.12 *Let A in $\mathcal{L}(H)$ be positive. Then:*

(a) *if there exists a basis of H consisting of eigenvectors of A then for all normal partial isometries U*

$$\|A - E_{\frac{1}{2}}\| \le \|A - U\|$$

where $E_{\frac{1}{2}}$ is as in Definition 6.3.8;

(b) *under the hypothesis of (a) if, further, $A - U \in \mathcal{C}_1$, then*

$$\|A - E_{\frac{1}{2}}\|_1 \le \|A - U\|_1;$$

(c) *for all normal partial isometries U such that $A - U \in \mathcal{C}_p$, where $1 < p < \infty$, there exists a basis of H consisting of eigenvectors of A and*

$$\|A - E_{\frac{1}{2}}\|_p \le \|A - U\|_p; \tag{6.3.24}$$

further, for strictly positive A such that $\frac{1}{2} \notin \sigma_p(A)$, equality occurs in (6.3.24) if and only if $U = E_{\frac{1}{2}}$.

Proof

(a) If $\{\phi_i\}$ is a basis of H consisting of eigenvectors of A, with $A\phi_i = \alpha_i \phi_i$ where $\alpha_i \ge 0$, then, with F as in Theorem 6.3.11, $F(A)\phi_i = F(\alpha_i)\phi_i = E_{\frac{1}{2}}\phi_i$ and hence $F(A) = E_{\frac{1}{2}}$. The result now follows from Theorem 6.3.11(a).

(b) As in (a), $F(A) = E_{\frac{1}{2}}$. The result now follows from Theorem 6.3.11(b).

(c) By Theorem 6.3.11(b) the map $F_p : U \mapsto \|A - U\|_p$, for $1 \le p < \infty$, also attains a global minimum. The result now follows, for $1 < p < \infty$, from Theorem 6.3.9.

□

Observe that we cannot deduce from Theorem 6.3.12 an inequality like (6.3.20) dealing with approximation to non-positive A (because, in the notation of (6.3.20), the partial isometries $\hat{U}_0^* U$ need to be normal).

In the light of Theorem 6.3.10, Theorem 6.3.12 raises the following question: in the infinite-dimensional case what happens if the partial isometries are not normal?

Exercises

1 Verify that V is a critical point of $U \mapsto \|A - U\|_2^2$ (for positive A and unitary U) if and only if $V = I$ using the formula, given by Theorem 2.4.4, for the derivative of $X \mapsto \|X\|_p^p$ when $1 < p < \infty$.

2 Aiken, Erdos and Goldstein's proof of the inequality (6.2.4) in Theorem 6.2.6 does not involve spectral approximants (when their paper was submitted, Bouldin's work on $\mathcal{C}_p$ spectral approximants had not yet appeared). Here is their reasoning [1, Theorem 3.5]:

(a) in finite dimensions the map $U \mapsto \|A - U\|_p$, for $1 < p < \infty$, has a local minimizer at $U = V$, say;
(b) $V = I$;
(c) for an arbitrary basis $\{\phi_i\}$ of the (now) infinite dimensional H let E_n be the projection onto $\mathcal{S}\{\phi_1, \ldots, \phi_n\}$ for each n so that

$$\|E_n A E_n - E_n\|_p \le \|E_n A E_n - E_n U E_n\|_p$$

and hence, on letting $n \to \infty$,

$$\|A - I\|_p \le \|A - U\|_p.$$

Fill in the gaps in the above reasoning and check that $\|E_n Q E_n\|_p \to \|Q\|_p$ as $n \to \infty$ if $Q \in \mathcal{C}_p$.

3 Construct an example such that the basis of H consisting of eigenvectors of the compact normal operator $A - E_U$ (for positive A and partially isometric U) is not a basis consisting of eigenvectors of A.

4 Verify Example 6.2.7 (Hint: if $s_1 \ge s_2 \ge 0$ are eigenvalues of $|A - U|$ show that

(a) $s_1^2 + s_2^2 = 2(2 + \alpha^2 - 2\cos\theta)$ and $s_1^2 s_2^2 = (\alpha^2 + 2\cos\theta - 2)$
(b) $s_1 + s_2 = 2\alpha$.

5 Prove Theorem 6.2.8.

6 Prove that if $A \in \mathcal{L}(H)$ and U varies over those partial isometrics such that $A - U \in \mathcal{C}_p$, where $1 < p < \infty$, then if V is a critical point of the map $F_p : U \to \|A - U\|_p^p$ it follows that $V^*A = AV^*$.

7 Verify Theorem 6.3.6(b).

8 Consider the following generalizations of the Lowdin orthogonalization.

(a) Let $\{f_1, \ldots, f_n\}$ be given basis of unit vectors of $\mathbb{C}^n$. It is required to replace $\{f_1, \ldots, f_n\}$ by an orthonormal basis $\{e_1, \ldots, e_n\}$ such that

  (i) $\{e_1, \ldots, e_n\}$ is close to $\{f_1, \ldots, f_n\}$,

(ii) $\{e_1, \ldots, e_n\}$ is close to some given orthonormal basis $\{g_1, \ldots, g_n\}$.

Let $Bf_i = e_i$, $1 \leq i \leq n$, and let $B = UL$ be the (unique) polar decomposition of B and let W be the unitary operator taking $\{Lf_1, \ldots, Lf_n\}$ to $\{g_1, \ldots, g_n\}$. For $0 \leq b \leq 1$, let $F_b(B)$ be defined by

$$F_b(B) = b \sum_{i=1}^{n} \|f_i - e_i\|^2 + (1 - b) \sum_{i=1}^{n} \|g_i - e_i\|^2.$$

Show that

$$F_b(B) = b\|L^{-1} - U\|_2^2 + (1 - b)\|W - U\|_2^2.$$

(b) More generally, let $A_1, \ldots A_m$ be in $\mathcal{L}(\mathbb{C}^2)$ and $b_1, \ldots, b_m$ be in $\mathbb{R}$ and such that $\sum_{i=1}^{m} b_i = 1$. For a unitary operator U let $G_b(U)$ be defined by

$$G_b(U) = \sum_{i=1}^{m} b_i \|A_i - U\|_2^2.$$

Show that minimizing $G_b(U)$ as U varies over the group of unitary operators in $\mathcal{L}(\mathbb{C}^n)$ is equivalent to maximizing the quantity $H_b(U)$ defined by

$$H_b(U) = \mathcal{R}\tau[U^*C] \text{ where } C = \sum_{i=1}^{m} b_i A_i.$$

Let $C = V|C|$ be the polar decomposition of C. Show that the maximum of $H_b(U)$ is $\|C\|$, and that this is attained whenever $U = V$ on the range of C.

Notes

A good presentation of the material in Sect. 6.1 is in [18], see also [2]. The proof of Theorem 6.1.1 is from [2, Theorem 2]. The material on isometric approximation is based on [1] (Lemma 6.2.2 is from [1, Theorem 3.2]). The content of Sect. 6.3 is based on [28] with Theorem 6.3.11(b) depending also on [34, Corollary 3].

Exercises: for Exercises 2, 4 and 5 see [1, Theorem 3.5, Example 3.8, Theorem 3.7], respectively; for Exercise 8 see [18, Sect. 5].

Bibliography

1. J.G. Aiken, J.A. Erdos, J.A. Goldstein, Unitary approximation of positive operators. Ill. J. Math. **24**, 61–72 (1980)
2. J.G. Aikten, J.A. Erdos, J.A. Goldstein, On Lowdin orthogonalization. Int. J. Quantum Chem. **18**, 1101–1108 (1980)
3. N.I. Akheiser, I.M. Glazman, in *Theory of Linear Operators in Hilbert Space*, vol. II (Unger, New York, 1963)
4. T.L. Allen, Bond energies and the interactions between next-nearest neighbours. I Saturated hydrocarbones diamond, sulphanes, 58 and organic sulfur components. J. Chem. Phys. **31**, 1039–1049 (1959)
5. J. Anderson, On normal derivations. Proc. Am. Math. Soc. **38**, 135–140 (1973)
6. J. Anderson, C. Foias, Properties which normal operators share with derivations and related operators. Pac. J. Math. **61**, 313–325 (1975)
7. S.K. Berberian, The Weyl spectrum of an operator. Indiana Univ. Math. J. **20**, 529–544 (1970)
8. R. Bhatia, Some inequalities for norm ideals. Commun. Math. Phys. **111**, 33–39 (1987)
9. R. Bhatia, F. Kittaneh, Approximation by positive operators. Linear Algebra Appl. **161**, 1–9 (1992)
10. S. Bouali, S. Cherki, Approximation by generalized commutators. Acta Sci. Math. (Szeged) **63**, 272–278 (1997)
11. R. Bouldin, Best approximation of a normal operator in the Schatten p-norm. Proc. Am. Math. Soc. **80**, 277–282 (1980)
12. J. Diestel, *Geometry of Banach Spaces*. Lecture Notes in Mathematics, vol. 485 (Springer, Berlin, 1975)
13. H. Dunford, J.T. Schwartz, *Linear Operators, Part II* (Interscience, New York, 1963)
14. M.J. Dupre, J.A. Goldstein, M. Levy, The nearest self-adjoint operator. J. Chem. Phys. **72**(1), 780–781 (1980)
15. J.A. Erdos, On the trace of a trace class operator. Bull. Lond. Math. Soc. **6**, 47–50 (1974)
16. K. Fan, A.J. Hoffman, Some metric inequalities in the space of matrices. Proc. Am. Math. Soc. **6**, 111–116 (1955)
17. J.A. Goldstein, M. Levy, Hilbert–Schmidt approximation problems arising in quantum chemistry. Adv. Appl. Math. **5**, 216–225 (1984)
18. J.A. Goldstein, M. Levy, Linear algebra and quantum chemistry. Am. Math. Mon. **98**, 710–718 (1991)
19. P.R. Halmos, Commutators of operators, II. Am. J. Math. **76**, 191–198 (1954)
20. P.R. Halmos, What does the spectral theorem say? Am. Math. Mon. **70**, 241–247 (1963)
21. P.R. Halmos, Positive approximants of operators. Indiana Univ. Math. J. **21**, 951–960 (1972)

© Springer International Publishing AG 2017

P.J. Maher, *Operator Approximant Problems Arising from Quantum Theory*,
DOI 10.1007/978-3-319-61170-9

22. P.R. Halmos, Spectral approximants of normal operators. Proc. Edin. Math. Soc. **19**, 51–58 (1974)
23. P.R. Halmos, *A Hilbert Space Problem Book*, 2nd edn. (Springer, New York, 1974)
24. P.R. Halmos, J.E. McLaughlin, Partial isometries. Pac. J. Math. **13**, 585–596 (1963)
25. R. Khalil, P.J. Maher, Spectral approximation in $L(H)$. Numer. Funct. Anal. Optim. **21**(5/6), 693–713 (2000)
26. D.C. Kleinecke, On operator commutators. Proc. Am. Math. Soc. **8**, 535–536 (1957)
27. M. Levy, W.J. Stevens, H. Schull, S. Hagstrom, Transferability of electron pairs between H_2O and H_2O_2. J. Chem. Phys. **61**, 1844–1856 (1974)
28. P.J. Maher, Partially isometric approximation of positive operators. Ill. J. Math. **33**, 227–243 (1989)
29. P.J. Maher, Some operator inequalities concerning generalized inverses. Ill. J. Math. **34**, 503–514 (1990)
30. P.J. Maher, The nearest self-adjoint operator. J. Chem. Phys. **92**(11), 6978 (1990)
31. P.J. Maher, Some norm inequalities concerning generalized inverses. Linear Algebra Appl. **174**, 99–110 (1992)
32. P.J. Maher, Commutator approximants. Proc. Am. Math. Soc. **115**, 995–1000 (1992)
33. P.J. Maher, Matrix approximation problems arising from quantum chemistry. Proc. Indian Natl. Sci. Acad. **64**, 715–723 (1998)
34. P.J. Maher, Spectral approximants concerning balanced and convex sets. Ann. Univ. Sci. Budapest. **46**, 177–181 (2003)
35. P.J. Maher, Self-commutator approximants. Proc. Am. Math. Soc. **134**, 157–165 (2007)
36. P.J. Maher, Commutator and self-commutator approximants, II. Filomat **24**(4), 1–7 (2010). www.pmf.ni.ac,yu/sajt/publiRacije/publiKacije–pocetna
37. C.A. McCarthy, C_p. Isr. J. Math. **5**, 249–271 (1967)
38. E.W. Packel, *Functional Analysis: A Short Course* (Intertext, New York, 1974)
39. H. Radjavi, P. Rosenthal, *Invariant Subspaces* (Springer, Berlin, 1973)
40. J.R. Ringrose, *Compact Non-Self-Adjoint Operators* (Van Nostrand Rheinhold, London, 1971)
41. D.D. Rogers, Approximation by unitary and essentially unitary operators. Acta Sci. Math. **39**, 141–151 (1977)
42. P.V. Shirokov, Proof of a conjecture of Kaplansky. Usp. Mat. Nauk **11**, 161–168 (1956)
43. B. Simon, *Trace Ideals and Their Applications* (Cambridge University Press, Cambridge, 1979)
44. H. Wielandt, Ueber die Unbeschränktheit der Operatoren des Quantenmechanik. Math. Ann. **121**, 21 (1949)
45. A. Wintner, The unboundedness of quantum-mechanical matrices. Phys. Rev. **71**, 738–739 (1947)

Index

© Springer International Publishing AG 2017
P.J. Maher, *Operator Approximant Problems Arising from Quantum Theory*,
DOI 10.1007/978-3-319-61170-9

Printed in the United States
By Bookmasters